# TABLE OF CONTENTS

# GUIDE TO SOLUTIONS

# FOR

# INORGANIC CHEMISTRY

**The Bedford Sixth Form**

**Learning Resources Centre**

**01234 291230**

learningresources@bedfordsixthform.ac.uk

*Oxford University Press, Walton Street, Oxford* OX2 6DP
*Oxford New York Toronto*
*Delhi Bombay Calcutta Madras Karachi*
*Petaling Jaya Singapore Hong Kong Tokyo*
*Nairobi Dar es Salaam Cape Town*
*Melbourne Auckland*
*and associated companies in*
*Berlin Ibadan*

*Oxford is a trade mark of Oxford University Press*

© *Steven H. Strauss, 1990*

*British Library Cataloguing in Publication Data*
*Data available*
*ISBN 0–19–855562–8*

*Printed in the United States of America*

*for Susie*

*and*

*my past, present, and future students*

# PREFACE

This book covers all of the in-chapter and end-of-chapter exercises in *Inorganic Chemistry* by Shriver, Atkins, and Langford. It is written to help you study a fascinating but intricate subject. As you read each chapter in *Inorganic Chemistry* and work on a set of exercises, you should consult the short answers given in the text. That will allow you to check quickly the overall accuracy of your answers. Then, *Guide to Solutions* will help you advance the learning process several steps further, by presenting complete solutions that permit a detailed understanding of the questions and the short answers and, more importantly, by explicitly tying the exercises and solutions to principles developed in the text. The solutions include virtually all of the figures and drawings asked for in the exercises. They also include many other figures that will help you visualize new concepts.

Many of the solutions will refer you to a particular chapter, section of a chapter, example, exercise, figure, table, or structural drawing in *Inorganic Chemistry*. These will always be clearly identified using boldface type. For example, the solution to **End-of-Chapter Exercise 1.1** will refer you to **Figure 1.4** in the text.

Learning about inorganic chemistry will require more than just a quick read and a glance at the short answers. There is no substitute for reading and re-reading the relevant material. It may surprise you to learn that most professors read the text more than once in the course of preparing lectures and exams. When studying for an exam, you should work through the exercises for a second time. If your solution is not as complete as the one in this book, you should review the relevant sections of the text. Keep in mind that one way to be sure that you understand something is to be able to explain it to someone else.

S. H. Strauss
Fort Collins, CO
October 1989

# ACKNOWLEDGMENT

I would like to thank those who reviewed various parts of this book, all of whom made valuable suggestions for its improvement: Drs. P. W. Atkins (University of Oxford), J. Evans (University of Southampton), W. Levason (University of Southampton), and D. W. H. Rankin (University of Edinburgh); Professors C. H. Langford (Concordia University), D. A. Phillips (Wabash College), D. M. Roddick (University of Wyoming), and D. F. Shriver (Northwestern University). I also thank Professors O. P. Anderson and A. K. Rappé (Colorado State University) for providing advice and assistance with the hardware and software used to produce the manuscript.

# CHAPTER 1

# ATOMIC STRUCTURE

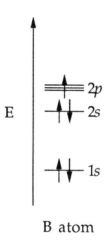

E

2*p*
2*s*

1*s*

B atom

Chemists frequently use energy level diagrams like this one, and they will be found throughout the text (e.g. **Figures 1.7, 1.8, 1.18**, and **1.23** in **Chapter 1**). The vertical axis is an energy axis, with increasing energy from bottom to top. The short horizontal lines represent the energies of orbitals, and the small arrows represent electrons in those orbitals. In this diagram, the electron configuration of a boron atom is shown pictorially.

## SOLUTIONS TO IN-CHAPTER EXERCISES

**1.1    How many orbitals in each subshell?** Each subshell is associated with an orbital angular momentum quantum number, $l$. A subshell contains $2 \cdot l + 1$ atomic orbitals. Thus, the $s$ subshell, with $l = 0$, consists of one orbital ($2 \times 0 + 1 = 1$); the $p$ subshell, with $l = 1$, consists of three orbitals ($2 \times 1 + 1 = 3$); the $d$ subshell, with $l = 2$, consists of five orbitals ($2 \times 2 + 1 = 5$); the $f$ subshell, with $l = 3$, consists of seven orbitals ($2 \times 3 + 1 = 7$); finally, the $g$ subshell, with $l = 4$, consists of nine orbitals ($2 \times 4 + 1 = 9$).

**1.2**    **Most probable distance for an electron in a 2s orbital.** The hydrogenic 2s wavefunction, $\psi$, is proportional to $(2 - Zr/a_0)e^{-Zr/2a_0}$. Since the most probable distance corresponds to the maximum in a plot of $r^2\psi^2$ vs. $r$, we must find the value(s) of $r$ for which $d(r^2\psi^2)/dr = 0$:

$$d(r^2\psi^2)/dr = 8r - 16Zr^2/a_0 + 8Z^2r^3/a_0^2 - Z^3r^4/a_0^3)e^{-Zr/a_0}$$

This expression is equal to 0 when $8 - 16Zr/a_0 + 8Z^2r^2/a_0^2 - Z^3r^3/a_0^3$ is equal to 0. Unlike the **Example**, this occurs for three different values of $r$. An obvious root is $r = 2a_0/Z$. If you factor out the obvious root, the following product of functions remains: $(2 - Zr/a_0)(4 - 6Zr/a_0 + Z^2r^2/a_0^2)$. The two roots of the quadratic expression can be found using the quadratic formula and are $r = 0.77a_0/Z$ and $5.2a_0/Z$. Substitution of the three roots into the expression for $r^2\psi^2$ will show that $r = 5.2a_0/Z$ gives the maximum value and hence is the most probable distance. The graphs below show plots of $r^2\psi^2$ vs. $r$ for hydrogenic 1s and 2s orbitals. While the plot for the 1s orbital has a single maximum, the 2s plot has two local maxima and a minimum, which correspond to the three solutions of $(8 - 16Zr/a_0 + 8Z^2r^2/a_0^2 - Z^3r^3/a_0^3) = 0$.

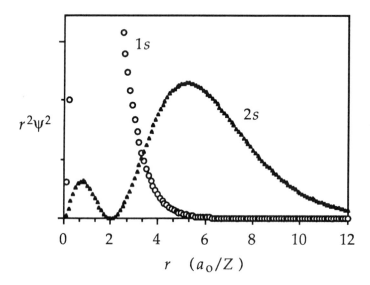

**1.3**    **Relative changes in $Z_{eff}$.** The configuration of the valence electrons, called the valence configuration, is as follows for the four atoms in question:

Li:     $2s^1$                    B:     $2s^22p^1$

Be:     $2s^2$                    C:     $2s^22p^2$

When an electron is added to the 2s orbital on going from Li to Be, $Z_{eff}$ increases by 0.63, but when an electron is added to an empty p orbital on going from B to C, $Z_{eff}$ increases by 0.72. The s electron already present in Li repels the incoming electron more strongly than the p electron already present in B repels the incoming p electron, because the incoming p electron goes into a new orbital. Therefore, $Z_{eff}$ increases by a smaller amount on going from Li to Be than from B to C. However, extreme caution must be exercised with arguments like this, because the effects of electron-electron repulsions are very subtle. This is illustrated in Period 3, where the effect is opposite to that just described for Period 2.

**1.4     Electron configuration of Ni and $Ni^{2+}$.** Following the **Example**, for an atom of Ni with $Z = 28$ the electron configuration is:

$$Ni:     1s^22s^22p^63s^23p^63d^84s^2  \text{ or } [Ar]3d^84s^2$$

Once again, the 4s electrons are listed last since the energy of the 4s orbital is higher than the energy of the 3d orbitals. Despite this ordering of the individual 3d and 4s energy levels for elements past Ca (see **Figure 1.19**), interelectronic repulsions prevent the configuration of a Ni atom from being $[Ar]3d^{10}$. For a $Ni^{2+}$ ion, with two fewer electrons than a Ni atom but with the same $Z$ as a Ni atom, interelectronic repulsions are less important; the higher energy 4s electrons are removed from Ni to form $Ni^{2+}$, and the electron configuration of the ion is:

$$Ni^{2+}:  1s^22s^22p^63s^23p^63d^8  \text{ or } [Ar]3d^8$$

**1.5     $I_1(Cl) < I_1(F)$?** When considering questions like these, it is always best to begin by writing down the electron configurations of the atoms or ions in question. If you do this routinely, a confusing comparison may become more transparent. In this case the relevant configurations are:

$$F \qquad 1s^2 2s^2 2p^5 \text{ or } [\text{He}]2s^2 2p^5$$

$$Cl \qquad 1s^2 2s^2 2p^6 3s^2 3p^5 \text{ or } [\text{Ar}]3s^2 3p^5$$

The electron removed during the ionization process is a $2p$ electron for F and a $3p$ electron for Cl. The principal quantum number, $n$, is lower for the electron removed from F ($n = 2$ for a $2p$ electron), so this electron is bound more strongly by the F nucleus than a $3p$ electron in Cl is bound by its nucleus.

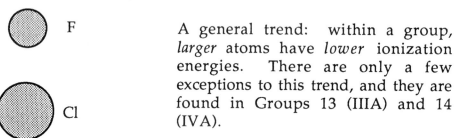

A general trend: within a group, *larger* atoms have *lower* ionization energies. There are only a few exceptions to this trend, and they are found in Groups 13 (IIIA) and 14 (IVA).

**1.6** $A_e(C) > A_e(N)$? The electron configurations of these two atoms are:

$$C: \quad [\text{He}]2s^2 2p^2 \qquad\qquad N: \quad [\text{He}]2s^2 2p^3$$

An additional electron can be added to the empty $2p$ orbital of C, and this is a favorable process ($A_e = 1.263$ eV). However, all of the $2p$ orbitals of N are already half occupied, so an additional electron added to N would experience sufficiently strong repulsions that the electron-gain process for N is unfavorable ($A_e = -0.07$ eV). This is despite the fact that the $2p$ $Z_{eff}$ for N is larger than $2p$ $Z_{eff}$ for C (see **Table 1.4**).

---

| SOLUTIONS TO END-OF-CHAPTER EXERCISES |
|---|

**1.1** **Draw the periodic table.** See **Figure 1.4** and the inside front cover of this book. You should start learning the names and positions of elements that you do not know. A blank periodic table can be found on the inside back cover of this book. You should make several photocopies of it and should test yourself from time to time.

**1.2     Relative abundances of H and He.** As discussed at the beginning of **Chapter 1**, the cosmic abundances of H and He today, which are very high relative to other elements, are not much different than only two hours after the Big Bang. The terrestrial abundances of these elements, however, are extremely small relative to, say, O, Mg, Si, and Fe. This is because H and He are the most volatile elements. They escaped the earth's gravitational field, which is relatively small, during a hot phase soon after the formation of the earth.

**1.3     Why is Fe abundant on earth?** There are two factors that combine to make Fe the most abundant element, by weight, on earth (by number of atoms, Fe is second only to O). The first factor is the high cosmic abundance of Fe (see **Figures 1.1** and **1.2**), which is related to its maximal nuclear stability. The second factor is the non-volatility of Fe. It condenses at temperatures below 1500 K, and so it was among the early condensates during the formation of the earth (see **Figure 1.3**). The elements with atomic numbers higher than iron's (Z = 26) are increasingly rare because they are built up from lighter elements by a variety of processes, including neutron capture. Their nuclei are unstable with respect to iron's, and so they are produced in non-equilibrium processes. In general, they cannot be more abundant than their lighter precursors.

**1.4     If λ = 0.25 Å for an electron, the potential difference = ?** The kinetic energy of the electron, $E_K$, after falling through the potential difference, is equal to the potential energy of the electron, $V$, before falling through the potential difference, and the potential energy is given by the product:

$$V = \text{(charge on the electron)(potential difference)}$$

The kinetic energy of the electron is related to the wavelength of the electron by the expressions:

$$E_K = p^2/2m_e \qquad \text{and} \qquad p = h/\lambda \qquad\qquad \textbf{Equation 1.1}$$

So $V = h^2/2\lambda^2 m_e = (6.626 \times 10^{-34}$ J s$)^2/2(2.5 \times 10^{-11}$ m$)^2(9.11 \times 10^{-31}$ kg$) =$ $3.9 \times 10^{-16}$ J.  The potential difference is obtained by dividing this result by one quantum of charge (the charge on the electron), $1.60 \times 10^{-19}$ C:

$$\text{potential difference} = (3.9 \times 10^{-16}\,\text{J})/(1.60 \times 10^{-19}\,\text{C}) = 2.4 \times 10^3\,\text{V}$$

**1.5    $E(\text{He}^+)/E(\text{Be}^{3+})$?** The ground state energy of a hydrogenic ion, like $\text{He}^+$ or $\text{Be}^{3+}$, is defined as the orbital energy of its single electron, which is given by **Equation. 1.5**:

$$E = -Z^2 m_e e^4/32\pi^2(\varepsilon_0)^2(h/2\pi)^2 n^2 \qquad (n = 1 \text{ for the ground state})$$

For the ratio $E(\text{He}^+)/E(\text{Be}^{3+})$, the constants can be ignored, and:

$$E(\text{He}^+)/E(\text{Be}^{3+}) = Z(\text{He}^+)^2/Z(\text{Be}^{3+})^2 = 2^2/4^2 = 0.25$$

**1.6    $E(\text{H}, n = 1) - E(\text{H}, n = 6)$.** The expression for $E$ given in **Equation 1.5** (see above) can be used for a hydrogen atom as well as for hydrogenic ions. The ratio $E(\text{H}, n = 1)/E(\text{H}, n = 6)$ can be determined as follows:

$$E(\text{H}, n = 1)/E(\text{H}, n = 6) = (1/1^2)/(1/6^2) = 36$$

Therefore, $E(\text{H}, n = 6) = (E(\text{H}, n = 1))/36 = -0.378$ eV, and the difference:

$$E(\text{H}, n = 1) - E(\text{H}, n = 6) = -13.2\,\text{eV}$$

**1.7    (a) Degenerate energy levels?** This means that the energy levels have the same energy.  An example of degenerate energy levels is a $p$ subshell of an atom;  for a given principal quantum number, $n$ (see part **(b)**, below) $p$ orbitals are said to be three-fold degenerate.

**(b)  Principal quantum number and its relation to $l$.** The principal quantum number, $n$, labels the shells of an atom.  For a hydrogen atom or a hydrogenic ion, $n$ alone determines the energy of all of the orbitals contained in a given shell (since there are $n^2$ orbitals in a shell, these would be $n^2$-fold degenerate. For a given value of $n$, the angular momentum quantum number, $l$, can assume all integer values from 0 to $n - 1$.

**1.8    Radial wavefunctions vs. radial distribution functions vs. angular wavefunctions.** The plots of $\psi$ vs. $r$ shown in **Figure 1.12** are plots of the radial parts of the total wavefunctions for the indicated orbitals. Notice that the plot of $\psi(2s)$ vs. $r$ takes on both positive (small $r$) and negative (large $r$) values, requiring that for some value of $r$ the wavefunction $\psi(2s) = 0$ (i.e. the wavefunction has a node at this value of $r$; for a hydrogen atom or a hydrogenic ion, $\psi(2s) = 0$ when $r = 2a_0/Z$ ). Notice also that the plot of $\psi(2p)$ vs. $r$ is positive for all values of $r$. Although a $2p$ orbital does have a node, it is not due to the radial wavefunction (the radial part of the total wavefunction). The plot of $4\pi r^2\psi^2$ vs. $r$ for a $1s$ orbital shown in **Figure 1.13** is a radial distribution function. A plot of $r^2\psi^2$ vs. $r$ for a $2s$ orbital is shown in the answer to **In-Chapter Exercise 2**, above. Whereas the radial distribution function for a $1s$ orbital has a single maximum, that for a $2s$ orbital has two maxima and a minimum (at $r = 2a_0/Z$ for hydrogenic $2s$ orbitals). The presence of the node at $r = 2a_0/Z$ for $\psi(2s)$ requires the presence of the two maxima and the minimum in the $2s$ radial distribution function. Using the same reasoning, the absence of a radial node for $\psi(2p)$ requires that the $2p$ radial distribution function has only a single maximum, as shown below:

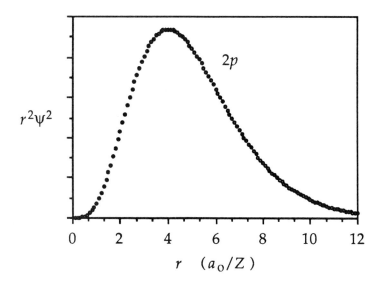

The angular wavefunctions are the familiar pictures that chemists draw to represent $s, p, d, f$, etc. orbitals, such as the ones in **Figures 1.14, 1.15,** and **1.16.** The familiar nodal plane for a $2p$ orbital is a property the orbital possesses because of the mathematical form of its angular wavefunction, not because of the mathematical form of its radial wavefunction.

**1.9     Penetration:** $E(4s) < E(3d)$**.** The higher the probability that an electron in an orbital will be found close to the nucleus, the more that electron *penetrates* the inner shells of electrons and the less completely that electron is shielded from the nucleus by the inner shells of electrons. A $4s$ orbital has a higher amplitude (higher value of $\psi$) close to the nucleus than does a $3d$ orbital. At the beginning of Period 4, this leads to the energy ordering $E(4s) < E(3d)$ (see **Figure 1.19**). However, the electrons that are added to the $3d$ subshell as we go from Sc to Zn can more effectively shield electrons in the $4s$ orbital than in a $3d$ orbital (in general, electrons with a given $n$ can more effectively shield other electrons with a higher value of $n$ than they can electrons with the same value of $n$). Therefore, from Sc on the energy ordering is $E(4s) > E(3d)$. A chemical consequence of this is that from Zn on, the $3d$ electrons are so tightly bound that they are not involved in the chemistry of the elements.

**1.10    $I_1$(Li).** The energy of an electron in a hydrogen atom or a hydrogenic ion is given by **Equation 1.5** (shown above in the answer to **End-of-Chapter Exercise 1.5**). This energy is equal to the negative of the ionization energy of the atom or ion, and it can be calculated for excited states by allowing $n$ to be greater than 1. Recalling that $I_1$ for a ground state H atom is 13.6 eV, the ionization energy of a $2s$ electron from H or from $Li^{2+}$ is given by:

$$E(2s) \text{ for H atom} = (-13.6 \text{ eV})(1/2^2)/(1/1^2) = -3.4 \text{ eV}$$

$$E(2s) \text{ for Li}^{2+} \text{ ion} = (-3.4 \text{ eV})(3^2/1^2) = -31 \text{ eV}$$

where the factors $2^2$ and $1^2$ in the first equation represent $n^2$ for a $2s$ orbital and a $1s$ orbital, respectively, while the factors $3^2$ and $1^2$ represent $Z^2$ for $Li^{2+}$ and H, respectively. From **Table 1.7** we see that $I_1$(Li) = 5.39 eV so $E(2s)$ for Li = $-5.39$ eV. The ionization energy of a ground state Li atom is larger than that of a H atom with its electron in its $2s$ orbital because, with three protons in its nucleus, the $2s$ orbital of Li experiences a larger $Z_{eff}$ than does the $2s$ orbital of H.  On the other hand, the ionization energy of a ground state Li atom is much smaller than that of a $Li^{2+}$ ion with its electron in its $2s$ orbital.  This is because the $2s$ orbital of Li experiences a much smaller $Z_{eff}$ than does the $2s$ orbital of $Li^{2+}$, since in the latter ion the $2s$ orbital is not shielded from the nucleus by the $1s$ electrons.  In summary:

$$Z(2s, Li^{2+}) = 3 > Z_{eff}(2s, Li) > Z(2s, H) = 1$$

**1.11** $I_2$ **of some period 4 elements.**  The second ionization energies of the elements Ca - Mn increase from left to right in the periodic table with the exception that $I_2(Cr) > I_2(Mn)$.  The electron configurations of the elements are:

| Ca | Sc | Ti | V | Cr | Mn |
|---|---|---|---|---|---|
| $[Ar]4s^2$ | $[Ar]3d^14s^2$ | $[Ar]3d^24s^2$ | $[Ar]3d^34s^2$ | $[Ar]3d^54s^1$ | $[Ar]3d^54s^2$ |

Both the first and the second ionization processes remove electrons from the $4s$ orbital of these atoms, with the exception of Cr.  The $4s$ electrons are poorly shielded by the $3d$ electrons, so $Z_{eff}(4s)$ increases from left to right and in general $I_2$ also increases from left to right.  While the $I_1$ process removes the sole $4s$ electron for Cr, the $I_2$ process must remove a $3d$ electron.  The higher value of $I_2$ for Cr relative to Mn is a consequence of the special stability of half–filled subshell configurations.

**1.12    Ground state electron configurations. (a) C.**  Four elements past He, which ends period 1, therefore $[He]2s^22p^2$.

(**b**) **F.**  Seven elements past He, therefore $[He]2s^22p^5$.

(**c**) **Ca.**  Two elements past Ar, which ends period 3 leaving the $3d$ subshell empty, therefore $[Ar]4s^2$.

(**d**) **Ga$^{3+}$.**  Thirteen elements but only ten electrons past Ar, and since it is a cation there is no doubt that $E(3d) < E(4s)$, therefore $[Ar]3d^{10}$.

(e) **Bi.** Twenty-nine elements past Xe, which ends period 5 leaving the 5d and the 4f subshells empty, therefore $[Xe]4f^{14}5d^{10}6s^26p^3$.

(f) **Pb$^{2+}$.** Twenty-eight elements but only twenty-six electrons past Xe, therefore $[Xe]4f^{14}5d^{10}6s^2$.

(g) **Nd.** Six elements past Xe, therefore $[Xe]4f^46s^2$.

(h) **U.** Six elements past Rn, which ends period 6 leaving the 6d and the 5f subshells empty (similar to Nd and its relation to Xe), therefore a reasonable configuration is $[Rn]5f^47s^2$. The actual electron configuration of U is $[Rn]5f^36d^17s^2$. The electron configurations of very heavy elements cannot be precisely predicted by a simple set of rules.

**1.13   More ground state electron configurations. (a) Sc.** Three elements past Ar, which ends Period 3 leaving the 3d subshell empty, therefore $[Ar]3d^14s^2$.

(b) **V$^{3+}$.** Five elements but only two electrons past Ar, and since it is a cation there is no doubt that $E(3d) < E(4s)$, therefore $[Ar]3d^2$.

(c) **Mn$^{2+}$.** Seven elements but only five electrons past Ar, therefore $[Ar]3d^5$.

(d) **Cr$^{2+}$.** Six elements but only four electrons past Ar, therefore $[Ar]3d^4$.

(e) **Co$^{3+}$.** Nine elements but only six electrons past Ar, therefore $[Ar]3d^6$.

(f) **Cr$^{6+}$.** Six elements past Ar, but with a +6 charge it has the *same* electron configuration as Ar, which is written as [Ar]. Sometimes inorganic chemists will write the electron configuration as $[Ar]3d^0$ to emphasize that there are no d electrons for this d-block metal ion.

**(g) Cu.** Eleven elements past Ar, but its electron configuration is not $[Ar]3d^9 4s^2$. The special stability experienced by completely filled subshells causes to actual electron configuration of Cu to be $[Ar]3d^{10}4s^1$.

**(h) Gd$^{3+}$.** Ten elements but only seven electrons past Xe, which ends Period 5 leaving the $5d$ and the $4f$ subshells empty, therefore $[Xe]4f^7$.

**1.14  $I_1$, $A_e$, and $\chi$ for Period 3.** The following values were taken from Tables **1.7**, **1.8**, and **1.9**:

| element | electron configuration | $I_1$ (eV) | $A_e$ (eV) | $\chi$ |
|---------|------------------------|------------|------------|--------|
| Na | $[Ne]3s^1$ | 5.14 | 0.548 | 0.93 |
| Mg | $[Ne]3s^2$ | 7.64 | −0.4 | 1.31 |
| Al | $[Ne]3s^2 3p^1$ | 5.98 | 0.441 | 1.61 |
| Si | $[Ne]3s^2 3p^2$ | 8.15 | 1.385 | 1.90 |
| P | $[Ne]3s^2 3p^3$ | 11.0 | 0.747 | 2.19 |
| S | $[Ne]3s^2 3p^4$ | 10.36 | 2.077 | 2.58 |
| Cl | $[Ne]3s^2 3p^5$ | 13.10 | 3.617 | 3.16 |
| Ar | $[Ne]3s^2 3p^6$ | 15.76 | −1.0 | — |

In general, $I_1$, $A_e$, and $\chi$ all increase from left to right across period 3 (or from top to bottom in the table above). All three quantities reflect how tightly an atom holds on to its electrons, or how tightly it holds on to additional electrons. The cause of the general increase across the period is the gradual increase in $Z_{eff}$, which itself is caused by the incomplete shielding of electrons of a given $n$ by electrons with the same $n$. The exceptions are explained as follows: $I_1(Mg) > I_1(Al)$ and $A_e(Na) > A_e(Al)$: both of these are due to the greater stability of $3s$ electrons relative to $3p$ electrons; $A_e(Mg)$ and $A_e(Ar) < 0$: filled subshells impart a special stability to an atom or ion — in these two cases the additional electron must be added to a higher energy subshell (for Mg) or shell (for Ar); $I_1(P) > I_1(S)$ and $A_e(Si) > A_e(P)$: the loss of an electron from S and the gain of an additional electron by Si both result in an ion with a half-filled $p$ subshell, which, like filled subshells, imparts a special stability to an atom or ion.

**1.15   $\eta$(Na) vs. $\eta$(Na$^+$).**  Hardness is defined by the expression:

$$\eta = 1/2(I - A_e)$$

**Tables 1.7** and **1.8** contain the relevant data.  For Na, $I_1$ = 5.14 eV and $A_e$ = 0.548 eV, so $\eta$ = 2.30 eV.  For Na$^+$, $I = I_2$(Na) = 47.29 eV and $A_e = -I_1$(Na) = –5.14 eV, so $\eta$ = 26.22 eV.  Therefore, $\eta$(Na$^+$) $>>$ $\eta$(Na).  The cause of this is primarily the difference $I_1$(Na$^+$) = $I_2$(Na) $>>$ $I_1$(Na).  An atom of Na is much softer than Na$^+$ because it outermost electron is in a $3s$ orbital and is much less tightly bound than the outermost electron of Na$^+$, which is in a $2p$ orbital.


**1.16   Correlation of size with $\chi$ and $\eta$.**  Absolute electronegativity and hardness can both be calculated from precisely measured atomic properties:

$$\chi = 1/2(I_1 + A_e) \qquad\qquad \eta = 1/2(I_1 - A_e)$$

In general, as size decreases, going up a group or from left to right across a period, $I_1$ and $A_e$ both increase.  Therefore, $\chi$ increases as size decreases.  However, the increases in $I_1$ are numerically greater than the increases in $A_e$.  For example, the increase in $I_1$ from Li (5.39 eV) to F (17.42 eV) is 12.03 eV, but the increase in $A_e$ from Li (0.618 eV) to F (3.399 eV) is only 2.781 eV.  As another example, the increase in $I_1$ from I (10.44 eV) to F is 6.98 eV, but the increase in $A_e$ from I (3.059 eV) to F is only 0.340 eV.  Due to the greater change in $I_1$ relative to $A_e$, $\eta$ *increases* as size decreases.

# CHAPTER 2

# MOLECULAR STRUCTURE

One of the most powerful tools of the inorganic chemist is group theory, a mathematical system for studying the symmetry of molecules and ions. One of the highest symmetry species possible is the octahedron, which possesses several 4-fold, 3-fold, and 2-fold proper rotation axes of symmetry, several mirror planes of symmetry, a center of symmetry, as well as 4-fold and 6-fold improper rotation axes of symmetry.

## SOLUTIONS TO IN-CHAPTER EXERCISES

**2.1    Lewis structure for PCl₃.** The four atoms supply $5 + 3 \cdot 7 = 26$ valence electrons. Since P is less electronegative than Cl, it is likely to be the central atom, so the 13 pairs of electrons are distributed as shown at the right. In this case, each atom obeys the octet rule. Whenever it is possible to follow the octet rule without violating other electron counting rules, we should do so.

13

**2.2    Resonance structures for $NO_2^-$.** First, draw a Lewis structure for the ion, trying to satisfy the octet rule. Then, permute any multiple bonds among equivalent atoms (i.e. those of the same chemical and structural type). In this case, the octet rule can be satisfied by drawing one N–O single bond and one N=O double bond. Two resonance structures, shown at the right, are necessary to depict the observation that both N–O bonds in the nitrite ion are equal in length and strength.

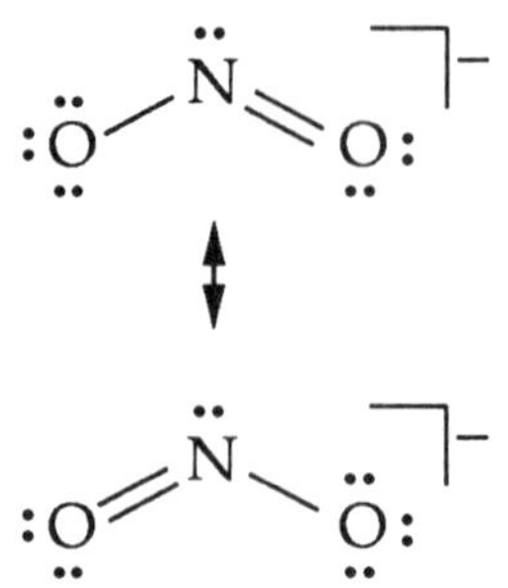

**2.3    Lewis structure for $XeO_6^{4-}$.** The seven atoms provide $8 + 6 \times 6 = 44$ valence electrons, so the –4 ion has a total of 48 electrons contained in 24 pairs of electrons. Since there are only 6 oxygen atoms, any Lewis structure must have four Xe–O single bonds and two Xe=O double bonds if the central Xe atom is to have 8 electron pairs. One of 15 possible resonance structures is shown at the right. There are 16 lone pairs of electrons on oxygen atoms, no lone pairs on the Xe atom, 4 single bonds (4 pairs of electrons) and 4 double bonds (4 pairs of electrons).

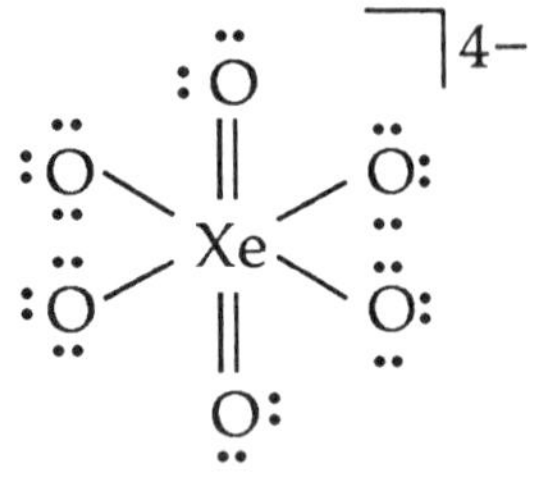

**2.4    Predict the shape of $XeF_4$.** The Lewis structure is shown below, accomodating an octet for the four F atoms and an expanded valence shell of 12 electrons for the Xe atom with the $8 + 4 \cdot 7 = 36$ valence electrons provided by the five atoms. The six pairs of electrons around the central Xe atom will will arrange themselves at the corners of an octahedron (like in $SF_6$). The two lone pairs will be at two opposite vertices of the octahedron, to minimize lone pair/lone pair repulsions. The resulting shape of the molecule, shown at the right, is square planar, with all of the F–Xe–F bond angles equal to 90°.

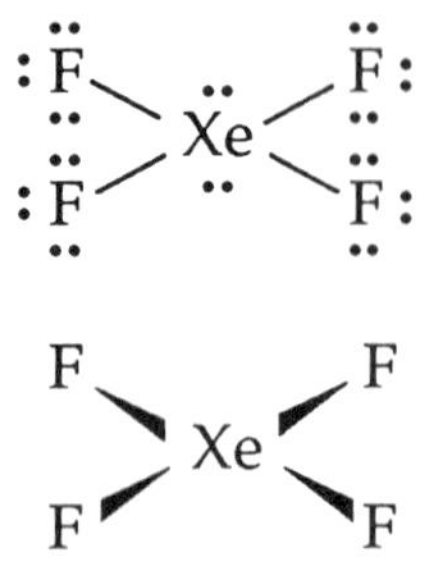

**2.5    Spectroscopic timescales.** As shown in the **Example**, two spectroscopic absorptions are indistinguishable if the lifetimes of the two states that give rise to the two different absorptions are *less than* $1/(2\pi\Delta v)$. Thus, if we are to distinguish two NMR absorptions (or resonances, as they are traditionally called) separated by 550 s$^{-1}$ (= 550 Hz), then the lifetimes of the two states must be *equal to or greater than* $1/(2\pi(550 \text{ s}^{-1})) = 2.89 \times 10^{-4}$ s.

**2.6    Identify the $C_3$ axes of an $NH_4^+$ ion.** Each of the N–H bond vectors corresponds to a $C_3$ axis for this ion, so there are four such axes. Since for a given $C_3$ axis the other three H atoms are related to each other by rotations of 120° or 240°, all four H atoms in the ammonium ion are symmetry related.

**2.7    (a) $BF_3$ point group?** All molecules possess the identity, $E$. Since $BF_3$ is *trigonal planar*, it is obvious that it possesses a 3–fold rotation axis ($C_3$) and a mirror plane of symmetry that coincides with the molecular plane ($\sigma_h$, since it is perpendicular to the $C_3$ axis, which is the "major axis"). There are also three 2–fold axes ($C_2$) that coincide with the three B–F bond vectors, and three more mirror planes, each of which contains a B–F bond and is perpendicular to $\sigma_h$ (these are called $\sigma_v$ since they are parallel to the major axis). The set of elements ($E, C_3, 3C_2, \sigma_h, 3\sigma_v$) corresponds to the group $D_{3h}$. Refer to **Table 2.7** and note that the $D_{3h}$ point group also contains an $S_3$ improper rotation axis. However, in many cases it is not necessary to find the complete set of symmetry elements to uniquely determine the point group of a molecule or ion. In fact, if we use the decision tree shown in **Figure 2.12** to determine the point group of $BF_3$, we find that the smaller set of elements ($C_3, 3C_2, \sigma_h$) uniquely corresponds to $D_{3h}$.

    **(b) $SO_4^{2-}$ point group?** Using the decision tree in **Figure 2.12** is generally the easiest way to determine the point group of a molecule or ion.

The sulfate ion (i) is non–linear, (ii) possesses four 3–fold axes ($C_3$), like $NH_4^+$ (see the answer to **In-Chapter Exercise 2.6**), and (iii) does not have a center of symmetry. The sequence of "no, yes, no" on the decision tree leads to the conclusion that $SO_4^{2-}$ belongs to the $T_d$ point group.

**2.8    Is ferrocene polar?** Like ruthenocene, the staggered conformation of ferrocene has a $C_5$ axis passing through the centroids of the $C_5H_5$ rings and the Fe atom. It also has five $C_2$ axes that pass through the Fe atom but are perpendicular to the major $C_5$ axis, so it belongs to one of the $D$ point groups, in this case $D_{5d}$ (it lacks the $\sigma_h$ plane of symmetry that a $D_{5h}$ molecule like ruthenocene possesses). Since the $D_{5d}$ conformation of ferrocene has a $C_5$ axis *and* perpendicular $C_2$ axes, it is not polar (see **Section 2.4**). You may find it difficult to find the n $C_2$ axes for a $D_{nd}$ structure. However, if you draw the mirror planes, the $C_2$ axes lie between them. In this case, one $C_2$ axis interchanges the front vertex of the top ring with one of the two front vertices of the bottom ring, while a second $C_2$ axis, rotated exactly 36° from the first one, interchanges the same vertex on top with the other front bottom one.

**2.9    Is the skew form of $H_2O_2$ chiral?** Except for the identity, $E$, the only element of symmetry that this conformation of hydrogen peroxide possesses is a $C_2$ axis that passes through the midpoint of the O–O bond and bisects the two O–O–H planes (these are *not* mirror planes of symmetry). Hence this form of $H_2O_2$ belongs to the $C_2$ point group, and it is chiral since this group does not contain any $S_n$ axes. In general, any structure that belongs to a $C_n$ or $D_n$ point group is chiral, as are molecules that are asymmetric ($C_1$ symmetry).

**2.10    Electron configurations for $S_2^{2-}$ and $Cl_2^-$.** The first of these two anions has the same Lewis structure as peroxide, $O_2^{2-}$. It also has a similar electron configuration to that of peroxide, except for the use of sulfur atom valence $3s$ and $3p$ atomic orbitals instead of oxygen atom $2s$ and $2p$ orbitals. There is no need to use sulfur atom $3d$ atomic orbitals, which are higher in energy than the $3s$ and $3p$ orbitals, since the $2·6 + 2 = 14$ valence electrons of $S_2^{2-}$ will not completely fill the stack of molecular orbitals constructed from sulfur atom

$3s$ and $3p$ atomic orbitals. Thus, the electron configuration of $S_2^{2-}$ is $1\sigma_g^2 2\sigma_u^2 3\sigma_g^2 1\pi_u^4 2\pi_g^4$. The $Cl_2^-$ anion contains one more electron than $S_2^{2-}$, so its electron configuration is $1\sigma_g^2 2\sigma_u^2 3\sigma_g^2 1\pi_u^4 2\pi_g^4 4\sigma_u^1$.

**2.11   Predict the photoelectron spectrum of SO.** Like carbon, sulfur is less electronegative than oxygen, and the MO energy level diagram for SO will resemble the diagram for CO shown in **Figure 2.29**, with S replacing C. Sulfur will use its valence $3s$ and $3p$ atomic orbitals instead of carbon's $2s$ and $2p$ orbitals, but the energy ordering should be about the same. We can even assume that the $3\sigma_g$ orbital for SO will be higher in energy than the $1\pi_u$ orbitals, as is the case for CO (see **Figure 2.29**), because S and C have about the same electronegativity (see **Table 1.9**). Since SO has two more electrons than CO, the electron configuration of SO is $1\sigma_g^2 2\sigma_u^2 1\pi_u^4 3\sigma_g^2 2\pi_g^2$, with two unpaired electrons in the $2\pi_g$ HOMOs (just like $O_2$). Therefore, the uv-PES spectrum of SO should resemble that shown for CO in **Figure 2.30**, with an additional series of peaks, corresponding to photoejection from the half-filled $2\pi_g$ level, at lower energies (i.e. to the left of) the intense σ peak.

**2.12   Estimate $\Delta H_f$ for $H_2S$.** We can "form" this molecule by considering the following reaction:

$$1/8\ S_8\ (g)\ +\ H_2\ (g)\ \rightarrow\ H_2S\ (g)$$

On the left side, we must break one H–H bond and also produce one sulfur atom from cyclic $S_8$. Since there are eight S–S bonds holding eight S atoms together, we must supply the mean S–S bond enthalpy *per S atom*. On the right side, we form two H–S bonds. From the values given in **Table 2.10**, we can estimate:

$$\Delta H_f \approx (436\ kJ\ mol^{-1}) + (264\ kJ\ mol^{-1}) - 2(338\ kJ\ mol^{-1}) = 24\ kJ\ mol^{-1}$$

This estimate indicates a slightly endothermic enthalpy of formation, but the experimental value, $-21\ kJ\ mol^{-1}$, is slightly exothermic.

**2.1 Lewis structures and VSEPR theory. (a) $GeCl_3^-$.** The Lewis structure is shown below. Only one resonance structure is important (each atom has an octet). Since the central Ge atom has four electron pairs, the structure of this ion is based on a tetrahedral array of electron pairs with one lone pair, and is described as a trigonal pyramid (like the structure of $NH_3$).

**(b) $SeF_4$.** Only one resonance structure is important for this molecule, and it is shown below. The F atoms each have an octet, while the Se atom has an expanded valence shell with ten electrons. Since the central atom has five electron pairs, the structure is based on a trigonal bipyramidal array of electron pairs with one lone pair (which is in the equatorial plane), and is described as a see–saw.

$$GeCl_3^- \qquad SeF_4 \qquad FCO_2^-$$

**(c) $FCO_2^-$.** The Lewis structure is shown above, with two resonance contributors to account for the equivalence of the two C–O bonds. Each atom has an octet. For the purposes of VSEPR analysis, the central C atom is treated as though it only had three electron pairs (see **Table 2.3** and the attendant discussion in the text), and so the geometry about the carbon atom is trigonal planar.

**(d) $CO_3^{2-}$.** Three resonance structures are necessary to account for the fact that the three C–O bonds in the carbonate anion are equivalent, and these are shown below. Each atom has an octet in all three resonance structures. The geometry about the central C atom is trigonal planar, for the same reason as described above for $FCO_2^-$.

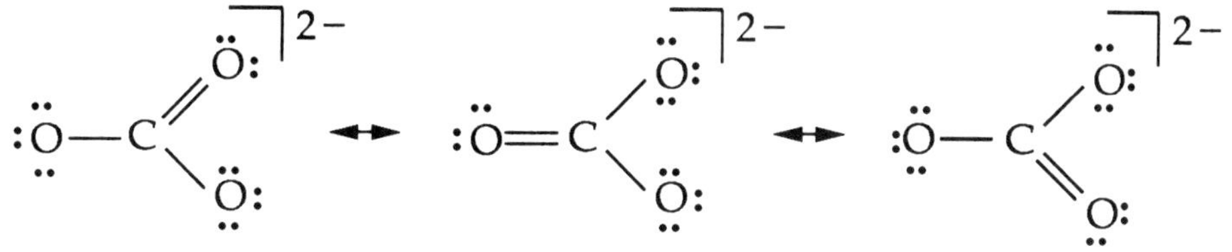

$$CO_3^{2-}$$

(e) **AlCl$_4^-$.** Only one resonace structure is necessary to achieve an octet around each atom and to account for the equivalence of the four Al–Cl bonds, and it is shown below. Based on four bonding electron pairs about the central Al atom, the structure of this anion is tetrahedral (like CCl$_4$).

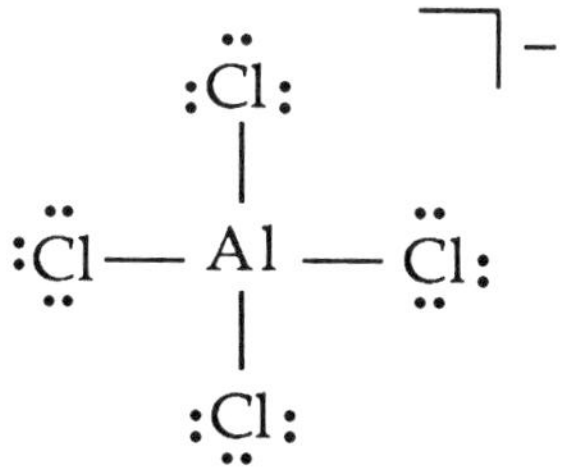

AlCl$_4^-$ 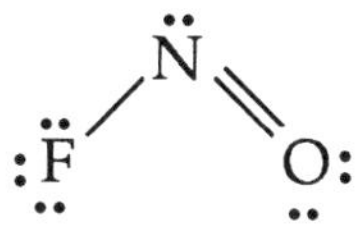                                                         FNO

(f) **FNO.** The least electronegative atom is likely to be the central atom. The Lewis structure of this molecule is shown above, requiring only one resonance structure to achieve an octet around all three atoms. For the purposes of VSEPR analysis, the central N atom is treated as though it only had three electron pairs, and since one of these is a lone pair, the geometry about the N atom is described as angular.

**2.2   More Lewis structures. (a) XeF$_4$.** The Lewis structure is shown below. Since the central Xe atom has six electron pairs, the geometry about it is based on an octahedral array of electron pairs with two lone pairs, which will occupy a pair of opposite vertices on the octahedron. Thus, the structure of this five-atom molecule is square planar.

XeF$_4$                              PF$_5$                              BrF$_3$

(b) **PF₅.** The Lewis structure, above, shows that the central P atom is surrounded by five bonding electron pairs. This leads to a trigonal bipyraminal geometry for PF$_5$, with two axial and three equatorial F atoms.

(c) **BrF₃.** The Lewis structure for this molecule is also shown above. Like the P atom in PF$_5$, the Br atom in BrF$_3$ is surrounded by five electron pairs, three bonding pairs and two lone pairs. The lone pairs will occupy equatorial sites in the trigonal bipyramidal array of electron pairs (see **Example 2.4**), leaving the four atoms of this molecule in the shape of a T. Note that the way the Lewis structure for BrF$_3$ has been drawn above is misleading in that it suggests that the lone pairs are axial and the three F atoms are equatorial. It is important to remember that Lewis structures do not normally convey structural information.

(d) **TeCl₄.** This molecule has an analogous Lewis structure (which is shown below) to that of SF$_4$, which is shown in **Example 2.4**. Thus, by similar reasoning to that given in the **Example**, TeCl$_4$ has a see-saw geometry.

TeCl$_4$                              ICl$_2^-$

(e) **ICl₂⁻.** The Lewis structure is shown above. Since the central I atom has five electron pairs, the structure is based on a trigonal bipyramidal array of electron pairs. Since the lone pairs will go in the equatorial plane of the trigonal bipyramid, the geometry about the I atom is linear.

**2.3    Sensitivity to lifetime broadening.** As discussed in **Section 2.2**, the energy of a state becomes less well defined as the lifetime of the state decreases.   In addition, the higher the frequency of the spectroscopic technique, the larger the frequency difference between two states of the molecule.  Thus, higher frequency techniques, such as ir spectroscopy ($\Delta v \approx$ (100 cm$^{-1}$)(3 x 10$^{10}$ cm s$^{-1}$) = 3 x 10$^{12}$ s$^{-1}$), are *less sensitive* to lifetime broadening than lower frequency techniques, such as NMR spectroscopy ($\Delta v \approx 10^3$ Hz = $10^3$ s$^{-1}$).  From **Table 2.5**, the order of increasing sensitivity to lifetime broadening is:

| Spectroscopic Technique | Approx. Frequency ($\approx 1/$(Timescale)) |
|---|---|
| X–ray PES | $10^{18}$ s$^{-1}$ |
| ultraviolet | $10^{15}$ s$^{-1}$ |
| visible | $10^{14}$ s$^{-1}$ |
| infrared | $10^{13}$ s$^{-1}$ |
| NMR | $10^1 - 10^9$ s$^{-1}$ |

**2.4    Calculate bond lengths. (a) CCl$_4$ (observed value = 1.77 Å).** From the covalent radii values given in **Table 2.9**, 0.77 Å for C and 0.99 Å for Cl, the C–Cl bond length in CCl$_4$ is predicted to be 0.77 Å + 0.99 Å = 1.76 Å.  The agreement with the experimentally observed value is excellent.

**(b) SiCl$_4$ (observed value = 2.01 Å).** The covalent radius for Si is 1.18 Å.  Therefore, the Si–Cl bond length in SiCl$_4$ is predicted to be 1.18 Å + 0.99 Å = 2.17 Å.  This is 8% longer than the observed bond length, so the agreement is not as good in this case.

**(c) GeCl$_4$ (observed value = 2.10 Å).** The covalent radius for Ge is 1.22 Å.  Therefore, the Ge-Cl bond length in GeCl$_4$ is predicted to be 2.21 Å.  This is 5% longer than the observed bond length.

**2.5**    **Draw sketches of $C_n$ axes and $\sigma$ planes.**    **(a)** NH$_3$. In the drawings below, the circle represents the nitrogen atom of ammonia and the diamonds represent the hydrogen atoms.  The mirror plane drawn is in the plane of the page, and it contains the nitrogen atom and the hydrogen atom on the left.  The other two hydrogen atoms are out of this plane, one above the page and one below.

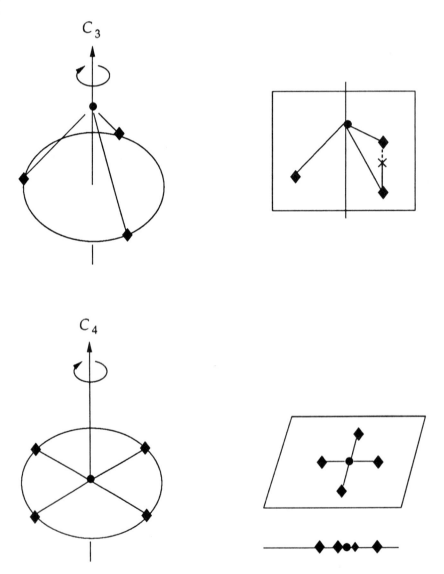

(b) The $PtCl_4^{2-}$ ion. In the drawings above, the circle represents the platinum atom of the tetrachloroplatinate anion and the diamonds represent the chlorine atoms. The mirror plane drawn is in the plane of the page, and it contains all five atoms. This plane is also drawn on its side, so that all five atoms seem to lie on a single line.

**2.6    $S_4$ or $i$? (a) $CO_2$.** This molecule has a center of inversion *and* an $S_4$ axis. The point group is $D_{\infty h}$, which includes all possible $C_n$ and $S_n$ axes.

(b) $C_2H_2$. This molecule also has $i$ and $S_4$. The point group is $D_{\infty h}$.

(c) $BF_3$. This molecule possesses neither $i$ nor $S_4$. It belongs to the $D_{3h}$ point group.

(d) $SO_4^{2-}$. This ion has three different $S_4$ axes, which are coincident with three $C_2$ axes, but there is no $i$. The point group is $T_d$.

**2.7    Assigning point groups. (a) $NH_2Cl$.** The only element of symmetry that this molecule possesses other than $E$ is a mirror plane that contains the N and Cl atoms and bisects the H–N–H bond angle. The set of symmetry elements $(E, \sigma)$ corresponds to the point group $C_s$ (the subscript comes from the German word for mirror, Spiegel).

(b) $CO_3^{2-}$. Carbonate anion is planar, so it possesses at least one plane of symmetry. Since this plane is perpendicular to the major proper rotation axis, $C_3$, it is called $\sigma_h$. In addition to the $C_3$ axis, there are three $C_2$ axes coinciding with the three C–O bond vectors. There are also other mirror planes and improper rotation axes, but the elements listed so far, $(E, C_3, \sigma_h, 3C_2)$, uniquely correspond to the $D_{3h}$ point group (note that $CO_3^{2-}$ has the same symmetry as $BF_3$). A complete list of symmetry elements is $E, C_3, 3C_2$, $S_3, \sigma_h$, and $3\sigma_v$.

**(c) SiF$_4$.** This molecule has four $C_3$ axes, one coinciding with each of the four Si–F bonds. In addition, there are six mirror planes of symmetry (any pair of F atoms and the central Si atom define a mirror plane, and there are always six ways to choose two objects out of a set of four). Furthermore, there is no center of symmetry. Thus, the set ($E$, $4C_3$, $6\sigma$, no $i$) describes this molecule and corresponds to the $T_d$ point group. A complete list of symmetry elements is $E$, $4C_3$, $3C_2$, $3S_4$, and $6\sigma_d$.

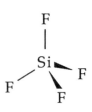

**(d) HCN.** Hydrogen cyanide is linear, so it belongs to either the $D_{\infty h}$ or the $C_{\infty v}$ point group. Since it does not possess a center of symmetry, which is a requirement for the $D_{\infty h}$ point group, it belongs to the $C_{\infty v}$ point group.

**(e) SiFClBrI.** This molecule does not possess any element of symmetry other than the identity element, $E$. Thus, it is asymmetric and belongs to the $C_1$ point group, the simplest possible point group.

**(f) BrF$_4^-$.** This anion is square planar. It has a $C_4$ axis and four perpendicular $C_2$ axes. It also has a $\sigma_h$ mirror plane. These symmetry elements uniquely correspond to the $D_{4h}$ points group. A complete list of symmetry elements is $E$, $C_4$, a parallel $C_2$, four perpendicular $C_2$, $S_4$, $i$, $\sigma_h$, $2\sigma_v$, and $2\sigma_d$.

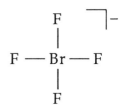

**2.8 The symmetry elements of orbitals. (a) An $s$ orbital.** An $s$ orbital, which has the shape of a sphere, possesses an infinite number of $C_n$ axes where n can be any number from 1 to $\infty$, plus an infinite number of mirror planes of symmetry. It also has a center of inversion, $i$. A sphere has the highest possible symmetry.

**(b) A $p$ orbital.** The + and − lobes of a $p$ orbital are not equivalent and therefore cannot be interchanged by potential elements of symmetry. Thus, a $p$ orbital does not possess a mirror plane of symmetry perpendicular to the long axis of the orbital. It does, however, possess an infinite number of mirror planes that pass through both lobes and include the long axis of the orbital. In addition, the long axis is a $C_n$ axis, where n can be any number from 1 to $\infty$ (in group theory this is referred to as a $C_\infty$ axis).

(c) A $d_{xy}$ orbital. The two pairs of + and − lobes of a $d_{xy}$ orbital are interchanged by the center of symmetry that this orbital possesses. It also possesses three mutually perpendicular $C_2$ axes, each one coincident with one of the three Cartesian coordinate axes. Furthermore, it possesses three mutually perpendicular mirror planes of symmetry, which are coincident with the the $xy$ plane and the two planes that are rotated by 45° about the z axis from the $xz$ plane and the $yz$ plane.

(d) A $d_{z2}$ orbital. Unlike a $p_z$ orbital, a $d_{z2}$ orbital has two large + lobes along its long axis, and a − torus (or doughnut) around the middle. In addition to the symmetry elements possessed by a $p$ orbital (see above), the infinite number of mirror planes that pass through both lobes and include the long axis of the orbital as well as the $C_\infty$ axis, a $d_{z2}$ orbital also possesses (i) a center of symmetry, (ii) a mirror plane that is perpendicular to the $C_\infty$ axis, (iii) an infinite number of $C_2$ axes that pass through the center of the orbital and are perpendicular to the $C_\infty$ axis, and (iv) an $S._\infty$ axis.

**2.9     Which species are polar?**  The sets of symmetry elements that independently require that a molecule is nonpolar are (i) a $C_n$ axis and a perpendicular $C_2$ axis (i.e., a $D$ point group), (ii) a $C_n$ axis and a $\sigma_h$ plane (i.e., a $C_{nh}$ point group), (iii) a center of inversion, $i$, or (iv) multiple $C_n$ axes with n > 2. The first two sets, (i) and (ii), were discussed in the text. An example of a molecule that possesses a center of symmetry but not a $C_n$ axis and a perpendicular $C_2$ axis is the planar conformation of $H_2O_2$ shown at the right. This molecule belongs to the $C_{2h}$ point group and is nonpolar. The fourth set, (iv) includes the tetrahedron, the octahedron, and the icosahedron. For these symmetries, there is no distinction between the x, y, and z directions (i.e. x, y, and z are triply degenerate).

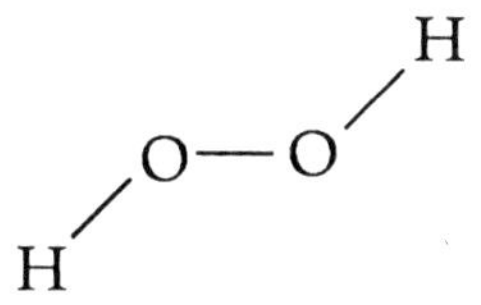

(a) $NH_2Cl$. This molecule does not meet any of the four criteria stated above, and so it is polar.

(b) $CO_3^{2-}$. The point group is $D_{3h}$, so on the basis of criterion (i), it is nonpolar.

(c) **SiF₄.** The point group is $T_d$, so on the basis of criterion (iv), it is nonpolar.

(d) **HCN.** This molecule does not meet any of the four criteria stated above, and so it is polar.

(e) **SiFClBrI.** This molecule too does not meet any of the four criteria stated above, and so it is polar.

(f) **BrF₄⁻.** This ion belongs to the $D_{4h}$ point group, so on the basis of criterion (i), it is nonpolar.


**2.10    Which species are chiral?** The symmetry criterion for chirality is the absence of an $S_n$ element of symmetry. Recall that $S_1 = \sigma$ and $S_2 = i$.

(a) **NH₂Cl.** This molecule does possess a mirror plane of symmetry, so it is not chiral.

(b) **CO₃²⁻.** The carbonate anion possesses four different planes of symmetry, so it is not chiral.

(c) **SiF₄.** This molecule, belonging to the $T_d$ point group, possesses six different planes of symmetry, so it is not chiral. Note that a tetrahedron also possesses three $S_4$ improper rotation axes that coincide with its $C_2$ axes.

(d) **HCN.** Since this molecule possesses an infinite number of mirror planes of symmetry, it is not chiral.

(e) **SiFClBrI.** This molecule does meet the criterion stated above for chirality, since it does not possess any $S_n$ element of symmetry (since it is $C_1$, or asymmetric, it does not possess *any* element of symmetry other than $E$).

(f) **BrF₄⁻.** This $D_{4h}$ ion possesses *many* $S_n$ elements of symmetry, including $\sigma_h$, $i$, $2\sigma_v$, $2\sigma_d$, and $S_4$. Therefore, it is not chiral.

**2.11   Writing electron configurations  (a)  Be$_2$.** Only four valence electrons for two Be atoms gives the electron configuration $1\sigma_g^2 2\sigma_u^2$.

(b) **B$_2$.** The electron configuration is $1\sigma_g^2 2\sigma_u^2 1\pi_u^2$.

(c) **C$_2^-$.** The electron configuration is $1\sigma_g^2 2\sigma_u^2 1\pi_u^4 3\sigma_g^1$.

(d) **F$_2^+$.** The electron configuration is $1\sigma_g^2 2\sigma_u^2 3\sigma_g^2 1\pi_u^4 2\pi_g^3$.

**2.12   How many unpaired electrons? (a)  O$_2^-$.** We must write the electron configurations for each species, using **Figure 2.22** as above, and then apply the Pauli exclusion principle to determine the situation for incompletely filled degenerate orbitals.   In this case the electron configuration is $1\sigma_g^2 2\sigma_u^2 3\sigma_g^2 1\pi_u^4 2\pi_g^3$.  With three electrons in the pair of $2\pi_g$ molecular orbitals, one electron must be unpaired.  Thus, the superoxide anion has a single unpaired electron.

(b) **O$_2^+$.** The configuration is $1\sigma_g^2 2\sigma_u^2 3\sigma_g^2 1\pi_u^4 2\pi_g^1$, so the oxygenyl cation also has a single unpaired electron.

(c) **BN.** We can assume that the energy of the $3\sigma_g$ molecular orbital is *higher* than the energy of the $1\pi_u$ orbitals, since that is the case for CO (see **Figure 2.29**).  Therefore, the configuration is $1\sigma_g^2 2\sigma_u^2 1\pi_u^4$, and, as observed, this diatomic molecule has no unpaired electrons.  If the configuration were $1\sigma_g^2 2\sigma_u^2 3\sigma_g^2 1\pi_u^2$, the molecule would have two unpaired electrons since each of the $1\pi_u$ orbitals would contain an unpaired electron, in accordance with the Pauli exclusion principle.

(d) **NO$^-$.** The exact ordering of the $3\sigma_g$ and $1\pi_u$ energy levels is not clear in this case, but it is not relevant either as far as the number of unpaired electrons is concerned.  The configuration is either $1\sigma_g^2 2\sigma_u^2 1\pi_u^4 3\sigma_g^2 2\pi_g^2$ or it is $1\sigma_g^2 2\sigma_u^2 3\sigma_g^2 1\pi_u^4 2\pi_g^2$.  In either case, this anion has one unpaired electron, and this electron occupies the set of antibonding $2\pi_g$ molecular orbitals.

**2.13   Determining bond orders.** The Lewis structures for the three species are shown below:

$$:\ddot{S}=\ddot{S}:\qquad :\ddot{C}l—\ddot{C}l:\qquad :\ddot{N}=\ddot{O}:\big]^{\,-}$$

(a) $S_2$.  The electron configuration of this diatomic molecule is $1\sigma_g^2 2\sigma_u^2 3\sigma_g^2 1\pi_u^4 2\pi_g^2$.  The bonding molecular orbitals are $1\sigma_g$, $1\pi_u$, and $3\sigma_g$, while the antibonding molecular orbitals are $2\sigma_u$, and $2\pi_g$.  Therefore, the bond order is $(1/2)((2 + 4 + 2) - (2 + 2)) = 2$, which is consistent with the double bond between the S atoms suggested by the Lewis structure.

(b) $Cl_2$.  The electron configuration is $1\sigma_g^2 2\sigma_u^2 3\sigma_g^2 1\pi_u^4 2\pi_g^4$.  The bonding and antibonding orbitals are the same as for $S_2$, above.  Therefore, the bond order is $(1/2)((2 + 4 + 2) - (2 + 4)) = 1$, which is in harmony with the single bond between the Cl atoms indicated by the Lewis structure.

(c) $NO^-$.  The electron configuration of $NO^-$, $1\sigma_g^2 2\sigma_u^2 1\pi_u^4 3\sigma_g^2 2\pi_g^2$, is the same as the configuration for $S_2$, shown above.  Thus, the bond order for $NO^-$ is 2, as for $S_2$, once again in harmony with the conclusion based on the Lewis structure.

**2.14   NMR interconversion.**  If we observe separate $^1H$ NMR signals, then the lifetimes of the two states of the molecule that give rise to the separate signals must be equal to or longer than $1/(2\pi\Delta v)$.  To find $\Delta v$ in this case, we recognize that 1 ppm at 90 MHz is exactly 90 Hz, so 2.00 ppm = 180 Hz = 180 $s^{-1}$.  Thus, the minimum lifetime is $1/(2\pi(180\ s^{-1})) = 8.84 \times 10^{-4}$ s.  The maximum rate at which the two states might be interconverting is equal to $1/(\text{minimum lifetime})$, which in this case is equal to $1/(8.84 \times 10^{-4}\ s) = 1.13 \times 10^3\ s^{-1}$ (see also **In-Chapter Exercise 2.5**).  The figure below shows the two extreme situations, called the fast exchange spectrum (the two states are rapidly interconverting and only one *sharp* signal is observed) and the slow exchange spectrum (the two states are slowly interconverting and a *sharp* signal for each state is observed).

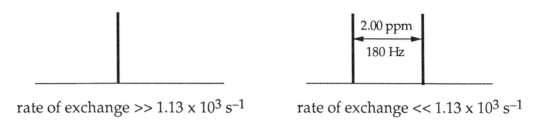

rate of exchange $\gg 1.13 \times 10^3$ s$^{-1}$          rate of exchange $\ll 1.13 \times 10^3$ s$^{-1}$

**2.15   The effect of ionization on bond orders.** If the ionization of a molecule *removes* an electron from a bonding molecular orbital, then the bond order of the resulting cation will be smaller than for the parent molecule (i.e. there will be fewer bonding electrons in the cation than in the neutral parent).   On the other hand, removal of an electron from an anti-bonding orbital will increase the bond order of the resulting cation relative to the neutral parent molecule.   By the same line of reasoning, *adding* an electron to a bonding orbital will increase the bond order, while *adding* an electron to an antibonding orbital will decrease the bond order.   In the case of the process $O_2 \rightarrow O_2^+ + e^-$, the electron is removed from the antibonding $\pi_g$ orbitals, so the bond order of the oxygenyl cation is higher than for $O_2$ (recall that the electron configuration of $O_2$ is $1\sigma_g{}^2 2\sigma_u{}^2 3\sigma_g{}^2 1\pi_u{}^4 2\pi_g{}^2$).   The $O_2^+$ O–O bond distance is therefore *shorter* than the $O_2$ O–O bond distance.   In the case of the process $N_2 + e^- \rightarrow N_2^-$, the electron is added to the antibonding $2\pi_g$ orbitals, so the bond order will decrease (the configuration of $N_2$ is $1\sigma_g{}^2 2\sigma_u{}^2 1\pi_u{}^4 3\sigma_g{}^2$).   The $N_2^-$ N–N bond distance is therefore *longer* than the $N_2$ N–N bond distance.   In the third case, $NO \rightarrow NO^+ + e^-$, the electron is removed from the antibonding $2\pi_g$ orbitals, so the bond order of $NO^+$ is higher than that of NO (the configuration of NO is $1\sigma_g{}^2 2\sigma_u{}^2 3\sigma_g{}^2 1\pi_u{}^4 2\pi_g{}^1$).   The $NO^+$ N–O bond distance is therefore *shorter* than the NO N–O bond distance.

**2.16   Calculating $\Delta H$ from mean bond enthalpies.** For the reaction:

$$2H_2 \text{ (g)} + O_2 \text{ (g)} \rightarrow 2H_2O \text{ (g)}$$

Since we must break two H–H bonds and one O=O bond on the left hand side of the equation and form four O–H bonds on the right hand side, the enthalpy change for the reaction can be estimated to be:

$$\Delta H \approx 2(436 \text{ kJ mol}^{-1}) + 497 \text{ kJ mol}^{-1} - 4(463 \text{ kJ mol}^{-1}) = -483 \text{ kJ mol}^{-1}$$

The experimental value is $-484$ kJ mol$^{-1}$, which is in closer agreement with the estimated value than ordinarily expected.

**2.17  Why is elemental nitrogen N$_2$ and elemental phosphorus P$_4$?**  Diatomic nitrogen has a triple bond holding the nitrogen atoms together, whereas six P–P single bonds hold together a molecule of P$_4$. If N$_2$ were to exist as N$_4$ molecules with the P$_4$ structure, then two N≡N triple bonds would be traded for six N–N single bonds, which are intrinsically weak. The net enthalpy change can be estimated from the data in **Table 2.10** to be $2(945 \text{ kJ mol}^{-1}) - 6(163 \text{ kJ mol}^{-1}) = 912 \text{ kJ mol}^{-1}$, which indicates that the tetramerization of nitrogen is *very* unfavorable. On the other hand, multiple bonds between Period 3 and larger atoms are not as strong as two times the analogous single bond, so P$_2$ molecules, each with a P≡P triple bond, would not be as stable as P$_4$ molecules, containing only P–P single bonds. In this case, the net enthalpy change for $2P_2 \rightarrow P_4$ can be estimated to be $2(481 \text{ kJ mol}^{-1}) - 6(200 \text{ kJ mol}^{-1}) = -238 \text{ kJ mol}^{-1}$.

N$_2$

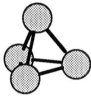

P$_4$

# CHAPTER 3

# POLYATOMIC MOLECULES AND SOLIDS

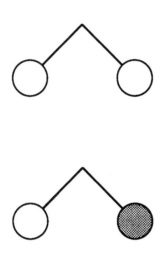

The H atom $1s$ orbitals in $H_2O$ form the two linear combinations shown. The unshaded lobes are + throughout and the shaded lobe is − throughout. These symmetry-adapted orbitals are labelled $a_1$ (top) and $b_2$ (bottom). The molecular orbitals of $H_2O$ are formed by making suitable linear combinations of these symmetry-adapted orbitals with O atomic orbitals of $a_1$ and $b_2$ symmetry.

---

SOLUTIONS TO IN-CHAPTER EXERCISES

---

**3.1    The $\pi$ MOs of an $E_8$ ring.** If we look down on the planar $E_8$ ring, we will see the tops of the atomic $p$ orbitals that compose the $\pi$–symmetry molecular orbitals. By extension of the ideas shown in **Figure 3.5**, the eight molecular orbitals are shown below. There are five different sets of orbitals (i.e. five different energy levels): orbital 1, which is singly degenerate and is the most stable orbital since it has no nodes perpendicular to the $E_8$ plane (all of these orbitals have a common nodal plane coincident with the plane of the molecule, as they must since they are composed of $p$ orbitals); orbitals 2a and 2b, which are doubly degenerate and next in energy with one node perpendicular to the $E_8$ plane; orbitals 3a and 3b, also doubly degenerate and next in energy with two nodes; orbitals 4a and 4b, the final set of doubly degenerate orbitals and next in energy with three nodes; and orbital 5, singly degenerate and highest in energy with four nodes perpendicular to the $E_8$ plane.

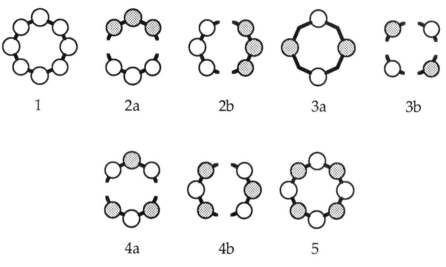

**3.2    What is the maximum possible degeneracy for an $O_h$ molecule?** This question, although asked about $SF_6$, could have been asked about any molecule with rigorous octahedral symmetry, i.e. a molecule that belongs to the $O_h$ point group. If we refer to the character table for this group, which is given in **Appendix 5**, we find that there are characters of 1, 2, *and* 3 in the column headed by the identity element, $E$. Therefore, the *maximum* possible degree of degeneracy of the orbitals in $SF_6$ is 3 (although non-degenerate and two-fold degenerate orbitals are allowed). As an example, the sulfur atom valence $p$ orbitals, and any molecular orbitals formed using them, are triply degenerate in $SF_6$.

**3.3    Orbital symmetry for a square planar array of H atoms.** We must adopt some conventions to answer this one. First, we assume that the combination of H atom 1s orbitals given looks like the figure shown below.

This array of H atoms has $D_{4h}$ symmetry. Inspection of the character table for this group, which is given in **Appendix 5**, reveals that there are three different types of $C_2$ rotation axes, i.e. there are three columns labelled $C_2$, $C_2'$, and $C_2''$. The first of these is the $C_2$ axis that is coincident with the $C_4$ axis; the second type, $C_2'$, represents two axes in the $H_4$ plane that do not pass through any H atoms; the third type, $C_2''$, represents two axes in the $H_4$ plane that pass through pairs of opposite H atoms.

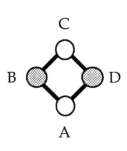

Now, instead of applying operations from all ten columns to this array, to see if it changes into itself (i.e. the +/− signs of the lobes stay the same) or if it changes sign, we can make use of a shortcut. Notice that the array changes into itself under the inversion operation through the center of symmetry. Thus the character for this operation, $i$, is 1. This means that the symmetry label for this array is one of the first four in the character table, $A_{1g}$, $A_{2g}$, $B_{1g}$, or $B_{2g}$. Notice also that for these four, the symmetry type is uniquely determined by the characters for the first five columns of operations, which are:

| $E$ | $C_4$ | $C_2$ | $C_2'$ | $C_2''$ |
|-----|-------|-------|--------|---------|
| 1 | −1 | 1 | −1 | 1 |

These match the characters of the $B_{2g}$ symmetry label.


**3.4  Orbital symmetry for a tetrahedral array of H atoms.** The molecule $CH_4$ has $T_d$ symmetry, and the combination of H atom $1s$ orbitals given also has $T_d$ symmetry. The group theory jargon that is used in this case is to say that the combination has the *total* symmetry of the molecule. This is true because each time the H atom array of orbitals is subjected to an operation in the $T_d$ point group, the array changes into itself. In each case, the character is 1. Each point group has one symmetry label (one row) for which all the characters are one, and for the $T_d$ point group it is called $A_1$ (see **Appendix 5**). Thus, the symmetry label of the given combination of H atom $1s$ orbitals is $A_1$.


**3.5  Is any Period 3 XH₂ molecule linear?** A Period 3 triatomic $XH_2$ molecule will have the same Walsh diagram as a Period 2 $XH_2$ molecule, shown in **Figure 3.17**, since the valence atomic orbitals used by a Period 3 central atom will be the same type (one $3s$ and three $3p$ orbitals) as those used by a Period 2 central atom (one $2s$ and three $2p$ orbitals). Reference to this Walsh diagram shows that an $XH_2$ molecule will only be linear if the $2a_1$ orbital is *not* occupied, and this occurs for $MgH_2$ (its electron configuration is $1a_1^2 1b_2^2$). Note that this prediction is only relevant for the *molecular* species $MgH_2$. The stable compound with this formula is actually a polymeric solid under normal conditions of temperature and pressure. In principle, a triatomic molecular species $NaH_2$ would also be linear, since with an electron configuration $1a_1^2 1b_2^1$, the all-important $2a_1$ orbital is empty.

**3.6    The conductivity behavior of VO. (a).** Since the conductivity of this compound increases with increasing temperature at low temperatures, it is a semiconductor at low temperatures.

**(b).** Above 125 K, the conductivity decreases gradually with increasing temperature, so the compound behaves like a metal at high temperatures. Note the abrupt rise in conductivity at about 125 K, which is the semiconductor/metal transition temperature. This compound, like many oxides with the formula MO, has the cubic NaCl structure, which will be discussed in **Chapter 4.** In the structure shown at the right, the $O^{2-}$ ions are the larger, open circles.

**3.7    The conductance of Ge at 370 K.** A plot of the data given in the **Example** is shown below (the squares are the data points). As predicted by **Equation 4**, the plot of ln $G$ vs. $1/T$ is linear. A linear least-squares fit to the data yields the equation ln $G = 11.1 - (4.24 \times 10^3)(1/T)$. At 370 K, we can interpolate ln $G$ from the graph (the circle) or calculate ln $G$ using the fitted equation. At this temperature, ln $G$ is found to be $-0.348$, so $G = 0.706$ S.

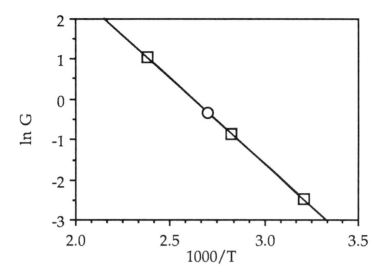

SOLUTIONS TO END-OF-CHAPTER EXERCISES

**3.1     Linear $H_4$ MOs.** Four atomic orbitals can yield four independent linear combinations.  The four relevant ones in this case, for a hypothetical linear $H_4$ molecule, are shown at the right in order of increasing energy.  The most stable orbital has the fewest nodes (i.e. the electrons in this orbital are not excluded from the internuclear regions), the next orbital in energy has only one node, and so on to the fourth and highest energy orbital, with three nodes (a node between each of the four H atoms).

**3.2     Point group and degenerate MOs for $SO_3^{2-}$.** Using the decision tree shown in **Figure 2.12**, we find the point group of this anion to be $C_{3v}$ (it is nonlinear, it only has one proper rotation axis, a $C_3$ axis, and it has three $\sigma_v$ mirror planes of symmetry).  Inspection of the $C_{3v}$ character table (**Table 3.1**) shows that the characters under

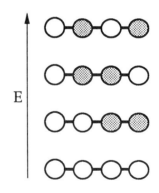

the column headed by the identity element, $E$, are 1 and 2. Therefore, the maximum degeneracy possible for molecular orbitals of this anion is 2. According to the character table, the S atom $3s$ and $3p_z$ orbitals are each singly degenerate (and belong to the $A_1$ symmetry type), but the $3p_x$ and $3p_y$ orbitals are doubly degenerate (and belong to the E symmetry type). Thus, the $3p_x$ and $3p_y$ atomic orbitals on sulfur can contribute to molecular orbitals that are two-fold degenerate.

**3.3 Point group and degenerate MOs of PF$_5$.** As above, we use the decision tree to assign the point group, in this case concluding that PF$_5$ has $D_{3h}$ symmetry (it has a trigonal bipyramidal structure, by analogy with PCl$_5$ (see **Table 2.2**); it is nonlinear, has only one high–order proper rotation axis, a $C_3$ axis, it has three $C_2$ axes that are perpendicular to the $C_3$ axis, and it has a $\sigma_h$ mirror plane of symmetry). Inspection of the $D_{3h}$ character table (**Appendix 5**) reveals that the characters under the $E$ column are 1 and 2, so the maximum degeneracy possible for a molecule with this symmetry is 2. The P atom $3p_x$ and $3p_y$ atomic orbitals, which are doubly degenerate and are of the $E'$ symmetry type (i.e. they *have* $E'$ symmetry), can contribute to molecular orbitals that are two-fold degenerate. In fact, if they contribute to molecular orbitals at all, they *must* contribute to two-fold degenerate ones.

**3.4 Symmetry-adapted orbitals for a square planar H$_4$ array.** The MH$_4$L$_2$ complex has $D_{4h}$ symmetry (a single $C_4$ axis, four $C_2$ axes perpendicular to it, and a $\sigma_h$ mirror plane of symmetry). Since our basis set of H atom orbitals consists of four H $1s$ orbitals (one per atom), we can construct exactly four symmetry–adapted combinations, which are shown below with their symmetry labels (the lines between the H atoms have no special meaning):

  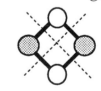

$A_{1g}$
(no nodes)
These two have $E_u$ symmetry
(one nodal plane)
$B_{1g}$
(two nodal planes)

(For a review of how to assign the symmetry label to a given orbital, see **In-Chapter Exercise 3.3**). The nodal planes are indicated by the dashed lines.

Note that the symmetry labels are written using capital letters. However, if we refer to the orbital with $A_{1g}$ symmetry, we call it an $a_{1g}$ orbital (this is also called the totally symmetric orbital). Note also that the nodal planes for the $e_u$ orbitals are perpendicular to each other — this is a necessary requirement for independent functions. The choice of $x$ and $y$ axes that led to the $B_{1g}$ assignment for the fourth orbital is to have them bisect the nodes. If we choose $x$ and $y$ to coincide with the two nodes, then the orbital will have $B_{2g}$ symmetry. Inspection of the $D_{4h}$ character table (**Appendix 5**) shows that $d_{z2}$ has $A_{1g}$ symmetry, so it can form molecular orbitals with the first of the H atom symmetry-adapted orbitals. There are no $d$ orbitals with $E_u$ symmetry, so no MOs can be formed using the H atom $e_u$ orbitals and metal $d$ orbitals. Finally, since $d_{x2-y2}$ has $B_{1g}$ symmetry, it can form MOs with the fourth H atom symmetry-adapted orbital.

**3.5    MOs for $H_2F^+$.** This ion has $D_{\infty h}$ symmetry. This point group is difficult to work with, but we can use a short cut by recognizing that the essential symmetry features of orbitals in this point group are $g$ (gerade, even to inversion) and $u$ (ungerade, odd to inversion). Furthermore, we must recognize that the F atom $2p_x$ and $2p_y$ atomic orbitals are orthogonal to the H atom $1s$ orbitals so they cannot mix with them.

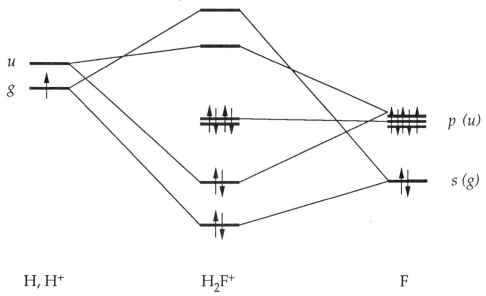

The two H atom symmetry-adapted orbitals are shown above and to the right. The MO diagram is also shown above.

**3.6    Average bond orders. (a) $NH_3$.** The molecular orbital energy diagram for ammonia is shown in **Figure 3.8**. The interpretation given in the text was that the $2a$ molecular orbital is almost nonbonding, so the electron configuration $1a^2 1e^4 2a^2$ results in only three bonds $((2 + 4)/2 = 3)$. Since there are three N–H "links," the average N–H bond order is 1 $(3/3 = 1)$.

**(b) $SF_6$.** The molecular orbital energy diagram for sulfur hexafluoride is shown in **Figure 3.9**. The configuration of this molecule is $1a^2 1t^6 1e^4$, and since the $e$ orbitals are nonbonding, there are only four bonds $((2 + 6)/2 = 4)$. Therefore, since there are six S–F "links," the average S–F bond order is 2/3 $(4/6 = 2/3)$.

**3.7    Describe the character of the HOMOs and LUMOs of $SF_6$.** The non-bonding $e$ HOMOs are pure F atom symmetry-adapted orbitals, and they do not have any S atom character whatsoever. They could only have S atom character if they were bonding or antibonding orbitals composed of atomic orbitals of both types of atoms in the molecule. On the other hand, the anti-bonding $t$ orbitals have both sulfur and fluorine character. Since sulfur is less electronegative than fluorine, its valence orbitals lie at higher energy than the valence orbitals of fluorine (from which the $t$ symmetry–adapted combinations were formed). Thus, the $t$ bonding orbitals lie closer in energy to the F atom $t$ combinations and hence they contain more F character (see **Figure 3.9**); the $t$ antibonding orbitals, the LUMOs, lie closer in energy to the S atom $3p$ orbitals and hence they contain more S character.

**3.8    Find isolobal fragments. (a).** The fragment $CH_3^-$ has a pyramidal structure with a single lone pair of electrons in the carbon atom. Ammonia, $NH_3$, has exactly the same molecular and electronic structure (see below), and hence is isolobal with $CH_3^-$.

**(b).** An O atom has six valence electrons, and in the ground state four of them are paired and two of them are unpaired in separate $p$ orbitals. An excited state of O has three pairs of electrons in three orbitals along with an empty orbital. The fragment $BH_3$ is isolobal with this state of an O atom.

(c). The fragment $Mn(CO)_5^-$ has many more valence electrons than any possible $NH_n$ fragment. Nevertheless, this metal carbonyl fragment has a single lone pair of electrons on the Mn atom, and hence is isolobal with $NH_3$ (see below).

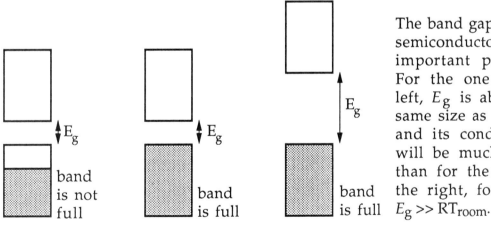

**3.9     Distinguish between a metal and a semiconductor. (a)** The band pictures for a metal and for two types of semiconductors are shown below:

The band gap, $E_g$, for semiconductors is an important property. For the one on the left, $E_g$ is about the same size as $RT_{room}$, and its conductivity will be much larger than for the one on the right, for which $E_g \gg RT_{room}$.

metal                    these two are semiconductors

**(b).** The electrical conductivity of a metal decreases as the temperature is increased. This is because increasing the temperature increases lattice vibrations, which reduce the freedom of electrons to move through the solid. In contrast, the electrical conductivity of a semiconductor increases as the temperature is increased, because increasing the temperature increases the thermal energy of the electrons in the solid, and more electrons are excited from the filled valence band to the empty conduction band. Thus, at higher temperatures, more charge carriers are present in a semiconductor and the conductivity increases. Over a broad temperature range, this thermal promotion of electrons has a greater influence on conductivity than the thermal vibrations which decrease conductivity.

(c). The conductivity of an insulator is very low but it does increases as the temperature is increased. Therefore, it is not possible to distinguish between a semiconductor and an insulator by the temperature dependence of their conductivities. In fact, the two types of materials are both thought of as semiconductors. They correspond to the two different situations shown in the band pictures above (the insulator has $E_g \gg RT_{room}$).

**3.10 Which are *n*-type and which are *p*-type semiconductors? (a) As-doped Ge.** Germanium has four valence electrons, but arsenic has five. Therefore, each As atom substituted for a Ge atom adds an electron to the lattice, so the doped material is *n*-type.

**(b) Ga-doped Ge.** Gallium, with three valence electrons, has one fewer than germanium. Therefore, each Ga atom substituted for a Ge atom removes an electron from the lattice, so the doped material is *p*-type.

**(c) Si-doped Ge.** Silicon and germanium both have four valence electrons, so substitution of Si for Ge does not change the total electron count of the lattice. Therefore, the doped material is neither *n*-type nor *p*-type. The dopant should have little effect on Ge.

**3.11 Calculate $E_g$ for TiO$_2$.** The band gap is equal to a change in energy, $\Delta E$, that is equivalent to a photon of light with a wavelength of 350 nm or less. Therefore:

$$E_g = \Delta E = h\nu = hc/\lambda = (6.626 \times 10^{-34} \text{ J s})(3.00 \times 10^8 \text{ m s}^{-1})/(350 \times 10^{-9} \text{ m})$$

$$E_g = 5.68 \times 10^{-19} \text{ J} = 3.55 \text{ eV} \qquad (1 \text{ eV} = 1.60 \times 10^{-19} \text{ J})$$

**3.12   Doping of $TiO_2$ with Ti(III).** Titanium(IV) has no valence electrons, but Ti(III) has one.  Therefore, each Ti(III) ion substituted for a Ti(IV) ion adds an electron to the lattice, so the doped material is *n*-type. We can see, in general, that doping a material with a lower oxidation state results in *n*-doping while doping with a higher oxidation state results in *p*-doping.

**3.13   Is GaAs-doped with Se *n*-doped or *p*-doped?** Selenium is electro-negative, like arsenic, and so will substitute for arsenic in the compound GaAs.  Since Se, with six valence electrons, has one more than As, which only has five, each Se atom substituted for a As atom adds an electron to the lattice, resulting in an *n*-type doped material.

**3.14   $\lambda_{max}$ for photoconduction in CdS.** As in **End-of-Chapter Exercise 3.11**, the band gap, in this case 2.4 eV, is equal to a change in energy that is equiv-alent to a photon of light with a wavelength of $\lambda$ or less.  Therefore:

$$E_g = \Delta E = h\nu = hc/\lambda$$

$$\lambda = hc/E_g = (6.626 \times 10^{-34} \text{ J s})(3.00 \times 10^8 \text{ m s}^{-1})/(2.4 \text{ eV})(1.60 \times 10^{-19} \text{ J eV}^{-1})$$

$$\lambda = 5.18 \times 10^{-7} \text{ m} = 518 \text{ nm}$$

Thus, light that has a wavelength of 518 nm *or less* will excite an electron from the valence band to the conduction band.  Light that has a wavelength of 518 nm has just enough energy to bridge the band gap, and light that has a shorter wavelength has more than enough energy.

**3.15   Calculate $\sigma(373 \text{ K})/\sigma(273 \text{ K})$ for Si.** From **Equation 3.4** we know that the conductivity of a semiconductor shows an Arrhenius-like temperature dependence:

$$\sigma = Ae^{-E_a/kT} \qquad \text{where} \qquad E_a \approx (1/2)E_g$$

Therefore:

$$\sigma(373\ K)/\sigma(273\ K) = e^{-(Ea/k)((1/373) - (1/273))} =$$

$$\exp(-((1.14\ eV)/2)(1.60 \times 10^{-19}\ J\ eV^{-1})(-9.82 \times 10^{-4}\ K^{-1})/(1.38 \times 10^{-23}\ J\ K^{-1})$$

$$\sigma(373\ K)/\sigma(273\ K) = e^{6.50} = 665$$

So, the conductivity of silicon increases by a factor of 665 between 0 °C and 100 °C. The band pictures below represent this behavior in general for an intrinsic semiconductor (the separations between shaded and light areas should be smooth curves). At temperature $T_1$, a few charge carriers have been promoted from the filled valence band into the empty conduction band, but at the higher temperature $T_2$, more charge carriers have been promoted and the conductivity is higher.

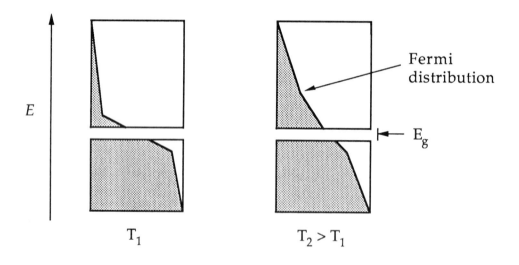

# CHAPTER 4

# THE STRUCTURES OF SOLIDS

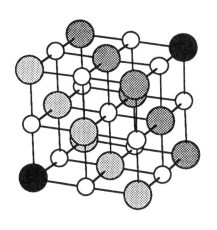

The rock-salt structure of NaCl is an example of an ionic solid built up from a close–packed array of anions. In the drawing at the left, the small open circles represent $Na^+$ ions. The larger circles represent $Cl^-$ ions, which are stacked together in a cubic close-packed (i.e. ABCABC . . .) array. The packing direction is along a body diagonal of the unit cell cube: the black $Cl^-$ ions are in A positions, the lightly shaded ones are in B positions, and the heavily shaded ones are in C positions. The $Na^+$ ions are in *all* of the octahedral holes formed by the $Cl^-$ ion array.

---

## SOLUTIONS TO IN-CHAPTER EXERCISES

**4.1    How many ions are in the NaCl unit cell?** The unit cell is shown in the drawing above. An ion that is completely inside of the cell counts as 1 ion in that cell, and the $Na^+$ ion in the center of the unit cell fits this description. An ion that is on a face of the unit cell is shared by two different unit cells and counts as 1/2 of an ion in each of the two cells. There are six $Cl^-$ ions in this category, one on each face of the unit cell cube (these are the nearest neighbors to the central $Na^+$ ion). An ion that is on an edge of the unit cell is shared by four different unit cells and counts as 1/4 of an ion in each of the four cells. Twelve $Na^+$ ions are in this category (these are the next nearest neighbors to the central $Na^+$ ion). Finally, an ion that is at a corner of the unit cell is shared by eight different unit cells and counts as 1/8 of an ion in each of the eight cells. There are eight $Cl^-$ ions in this category. Therefore, the total count is as follows:

| Inside: | 1 | Counts: | 1 |
|---------|----|---------|---|
| Face:   | 6  |         | 3 |
| Edge:   | 12 |         | 3 |
| Corner: | 8  |         | 1 |
|         |    | Total:  | 8 |

As in the **Example**, the eight ions in the unit cell are four cations and four anions, since the rock-salt structure is only found for 1:1 salts.

**4.2    Predict the coordination number of RbCl using a structure map.** The electronegativities of Rb and Cl, taken from **Table 1.9**, are 0.82 and 3.16, respectively, so $\Delta\chi = 2.34$. Since rubidium is in Period 5 and chlorine is in Period 3, the average principal quantum number is 4. The point $(\Delta\chi, n) = (2.34, 4)$ is not on the structure map shown in **Figure 4.21**, but extrapolation leads to the conclusion that this point lies in the C.N. = 6 region of the map. Therefore, the rock-salt structure is predicted for RbCl, and this is in fact the observed structure at normal temperatures and pressures.

**4.3    Calculate $\Delta H_L$ for $MgBr_2$.** We proceed as in the **Example**, calculating the total enthalpy change for the Born-Haber cycle and setting it equal to $\Delta H_L$. In this case, it is important to recognize that two $Br^-$ ions are required, so the enthalpy changes for (i) vaporization of $Br_2$ (*l*) and (ii) breaking the Br–Br bond in $Br_2$ (*g*) are used without dividing by 2, as was done for KCl. Furthermore, the first *and* second ionization enthalpies for Mg (*g*) must be added together for the process Mg (*g*) $\rightarrow$ $Mg^{2+}$ (*g*) + 2e⁻. The Born-Haber cycle for $MgBr_2$ is shown below, with all of the enthalpy changes given in kJ mol⁻¹. These enthalpy changes are not to scale. The lattice enthalpy is equal to 2421 kJ mol⁻¹. Note that $MgBr_2$ is a stable compound despite the enormous enthalpy of ionization of magnesium — the very large lattice enthalpy more than compensates for this positive enthalpy term. Note the standard convention used: lattice enthalpies are positive enthalpy changes.

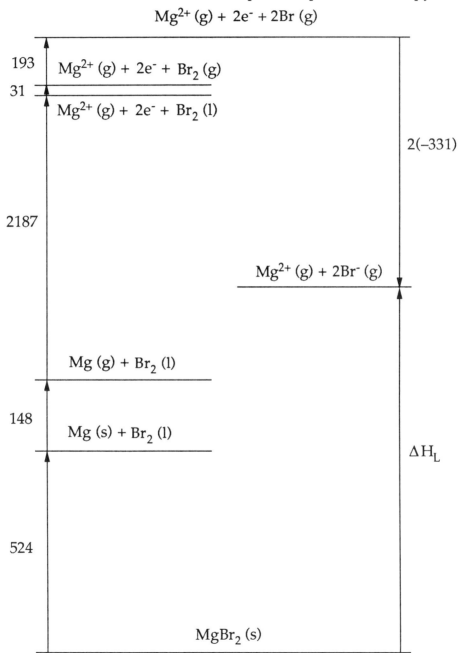

$Mg^{2+}$ (g) + 2e⁻ + 2Br (g)

193    $Mg^{2+}$ (g) + 2e⁻ + $Br_2$ (g)

31    $Mg^{2+}$ (g) + 2e⁻ + $Br_2$ (l)

2(−331)

2187

$Mg^{2+}$ (g) + 2Br⁻ (g)

Mg (g) + $Br_2$ (l)

148    Mg (s) + $Br_2$ (l)

$\Delta H_L$

524

$MgBr_2$ (s)

**4.4    Coulombic energy for CsCl.**  The coulombic energy for a given ionic substance is proportional to the Madelung constant, $A$, and inversely proportional to the interionic distance, $d$.  Values of Madelung constants are given in **Table 4.3** (rock-salt $A = 1.748$, cesium chloride $A = 1.763$), but $d$ must be calculated.   To do this we must find the ionic radius for $Cs^+$ with coordination number 6 (rock-salt structure) and coordination number 8 (cesium chloride structure) and add each of these to the ionic radius of $Cl^-$, which is 1.81 Å.  The radii for $Cs^+$ are found in **Table 4.2** and are 1.67 Å and 1.74 Å, respectively, for C.N. 6 and C.N. 8, so the two values of $d$ are 3.48 Å (C.N. 6, rock-salt structure) and 3.55 (C.N. 8, cesium chloride structure).  Therefore:

$$V(CsCl, \text{rock-salt structure}) \propto A/d = (1.748)/(3.48) = 0.502 \text{ Å}^{-1}$$

$$V(CsCl, \text{cesium chloride structure}) \propto A/d = (1.763)/(3.55) = 0.497 \text{ Å}^{-1}$$

Once again, the rock-salt structure is seen to be marginally more stable, in this case by only 1% $((0.502)/(0.497) = 1.01)$.  Since CsCl adopts the cesium chloride structure under normal conditions of temperature and pressure, this calculation of the coulombic contribution to the lattice energy is seen to be too simple to account for the observed structure.  Nevertheless, we can conclude that for CsCl the two structures may only differ in stability by a few kJ mol$^{-1}$.  In fact, above 469 °C, CsCl undergoes a phase transition and adopts the rock-salt structure.

**4.5    Is MgO properly regarded as an ionic solid?**  The experimental lattice enthalpy is 3.85 MJ mol$^{-1}$.  We must compare this result with the potential energy, $V$, calculated assuming an ionic model, using the Born-Mayer equation.   To use this equation, we must first calculate the interionic distance, $d$, for MgO.  Since this substance adopts the rock-salt structure, it is appropriate to use C.N. 6 ionic radii, which for $Mg^{2+}$ and $O^{2-}$ are 0.72 Å and 1.40 Å, respectively.  Thus, $d = 2.12$ Å and:

$$V = (1.39 \text{ MJ Å mol}^{-1})(4/2.12 \text{ Å})(1 - (0.345)/(2.12))(1.748)$$

$$V = 3.82 \text{ MJ mol}^{-1}$$

This calculated value of the potential energy is 99% of the experimentally determined lattice energy, so it is indeed proper to regard MgO as an ionic solid. Note the factor of 4 in the term (4/2.12 Å), which arises because $z_A z_B = 2 \times 2 = 4$. This factor of 4 is, to a good approximation, the only difference between $V$ for MgO and $V$ for LiF, for which $d = 2.09$ Å. In harmony with this, note that $\Delta H_L$ for LiF, 1037 kJ mol$^{-1}$ (from **Table 4.4**), is approximately 1/4 the value of $\Delta H_L$ for MgO.

**4.6    Calculate $\Delta H_L$ for CaSO$_4$ using the Kapustinskii equation.** To use the Kapustinskii equation, we note that the number of ions per formula unit, $n$, is 2, the charge numbers are 2+ for calcium ion and 2– for sulfate ion, and $d = 3.30$ Å (the C.N. 6 ionic radius for Ca$^{2+}$ is 1.00 Å (see **Table 4.2**) and the thermochemical radius of SO$_4^{2-}$ is 2.30 Å (see **Table 4.5**)). Therefore:

$$\Delta H_L = (8/3.30 \text{ Å})(1 - (0.345)/(3.30 \text{ Å}))(1.21 \text{ MJ mol}^{-1}) = 2.63 \text{ MJ mol}^{-1}$$

While this is a large value, it is considerably smaller than $\Delta H_L$ for MgO (another 2+/2– ionic solid), calculated in **In-Chapter Exercise 4.5**, above. This makes sense, because both the cation and anion are smaller in MgO than in CaSO$_4$.

**4.7    Which alkaline earth sulfate has the lowest decomposition temperature?** The enthalpy change for the reaction:

$$\text{MSO}_4 \ (s) \ \rightarrow \ \text{MO} \ (s) \ + \ \text{SO}_3 \ (g)$$

includes several terms, including the lattice enthalpy for MSO$_4$, the lattice enthalpy for MO, and the enthalpy change for removing O$^{2-}$ from SO$_4^{2-}$. The last of these is constant as we change M$^{2+}$ from Mg$^{2+}$ to Ba$^{2+}$, but the lattic enthalpies change considerably. The lattice enthalpies for MgSO$_4$ and MgO are both larger than those for BaSO$_4$ and BaO, simply because Mg$^{2+}$ is a smaller cation than Ba$^{2+}$. However, the *difference* between the lattice enthalpies for MgSO$_4$ and BaSO$_4$ is a smaller number than the difference between the lattice enthalpies for MgO and BaO (the larger the anion, the less changing the size of the cation affects $\Delta H_L$). Thus, going from MgSO$_4$ to MgO is thermodynamically more favorable than going from BaSO$_4$ to BaO, because the *change* in $\Delta H_L$ is greater for the former than for the latter. Therefore, magnesium sulfate will have the lowest decomposition temperature and barium sulfate the highest.

## SOLUTIONS TO END-OF-CHAPTER EXERCISES

**4.1    Which of the following are close packed? (a) ABCABC . . .** Any ordering scheme of planes is close-packed if no two adjacent planes have the same position (i.e. if no two planes are *in register*). When two planes are in register, the packing looks like the figure below and to the right, whereas the packing in a close-packed structure allows the atoms of one plane to fit more efficiently into the spaces between the atoms in an adjacent plane, like the figure below and to the left. Notice that the empty spaces between the atoms in the figure to the left are much smaller than in the figure to the right. The efficient packing exhibited by close-packed structures is why, for a given type of atom, close-packed structures are more dense than any other possible structure. In the case of an ABCABC . . . structure, no two adjacent planes are in register, so the ordering scheme is close-packed.

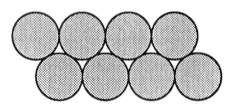  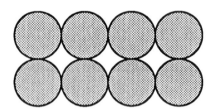

(b) **ABAC . . .** Once again, no two adjacent planes are in register, so the ordering scheme is close-packed.

(c) **ABBA . . .** The packing of planes using this sequence will put two B planes next to each other as well as two A planes next to each other, so the ordering scheme is not close-packed.

(d) **ABCBC . . .** No two adjacent planes are in register, so the ordering scheme is close-packed.

(e) **ABABC** . . . No two adjacent planes are in register, so the ordering scheme is close-packed.

(f) **ABCCBA** . . . The packing of planes using this sequence will put two C planes next to each other as well as two A planes next to each other, so the ordering scheme is not close-packed.

**4.2    Mark B and C positions in a layer of atoms at A positions.** In the drawing below, the spheres represent atoms in A positions, the ⊗ symbols are at B positions, and the O symbols are at C positions.

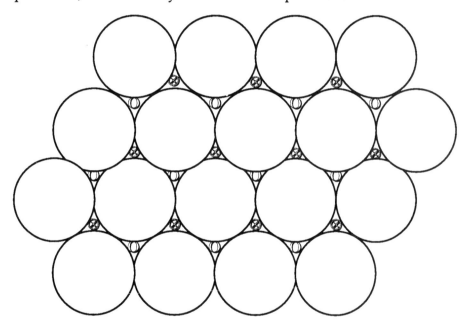

**4.3    Show that an octahedral hole in a close–packed lattice has a radius 0.414 times the radius of a packing sphere.** Using the hint, we draw a slice through an octahedral hole and four coplanar packing spheres. The drawing below shows this slice (the packing spheres above and below the octahedral

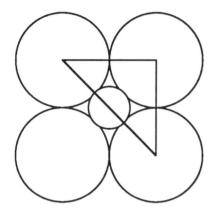

hole, which is the small sphere, are missing). If the radius of the packing spheres is set equal to 1, then the isosceles right triangle shown has sides of length 2 (twice the radius of a packing sphere) and a hypoteneuse of length $2\sqrt{2} = 2.828$. Since the hypoteneuse is equal to twice the radius of a packing sphere, which is 2, plus twice the radius of the octahedral hole, the radius of the octahedral hole is $(0.828)/2 = 0.414$.

**4.4    Describe the structure of WC.**  A solid-state structure is generally described, when appropriate, as a close-packed array of one type of atom or ion with other types of atoms or ions in the interstitial holes between the packing spheres.  Tungsten atoms are considerably larger than carbon atoms, so in this case it is best to envision WC as a close-packed lattice of W atoms with C atoms in interstitial holes.  Since the compound has the rock-salt structure, the structure is described as a *cubic* close-packed array of W atoms with C atoms in *all* of the *octahedral* holes.

**4.5    Is the alloy CuZn substitutional or interstitial?**  If CuZn has a CsCl structure, each metal is surrounded by eight nearest neighbor atoms of the other type.  Cu and Zn each have close-packed structures or structures that are nearly close-packed, so CuZn is neither substitutional nor interstitial.

**4.6    Which elements form truly ionic compounds?**  The ions that form solids that conform well to the ionic model are those that are small, hard, not polarizable, and not highly charged (i.e. they are +1, +2, −1, or −2).  In the drawing below, note that Be has been omitted, although all of the other +2 alkaline earths form solids that are well described by the ionic model.  This is because $Be^{2+}$ is *so* small that it is too polarizing, and compounds like $BeF_2$ and BeO have considerable covalent character.

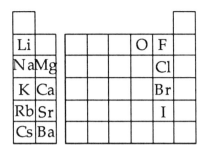

| Li | | | | | O | F | |
|---|---|---|---|---|---|---|---|
| Na | Mg | | | | | Cl | |
| K | Ca | | | | | Br | |
| Rb | Sr | | | | | I | |
| Cs | Ba | | | | | | |

**4.7    The rock-salt structure of NaCl.** The structure of NaCl is shown below. The $Na^+$ ions are the small open spheres while the $Cl^-$ ions are the larger shaded spheres. Each type of ion has six nearest neighbors which are the opposite type of ion. Each $Na^+$ ion has twelve next nearest neighbors, and these are $Na^+$ ions (one along each of the twelve edges of the cubic unit cell shown below). The six more heavily shaded $Cl^-$ ions are in a single "close-packed" plane (the $Cl^-$ ions cannot really be close-packed — if they were, the $Na^+$ ions would not fit into the octahedral holes). These have been redrawn in the plane of the page to the right of the unit cell. If we look down one of the four $C_3$ axes that the unit cell cube possesses, the x marks the spot where this axis intersects the close-packed plane. Note that the $Na^+$ ion that is at the center of the unit cell cube lies on the $C_3$ axis. This ion lies at position x but not in the plane of the page. Instead, it lies midway between this close-packed plane and an adjacent one.

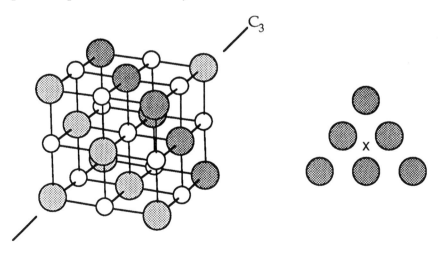

**4.8    The cesium chloride structure.** The unit cell for this structure is shown in **Figure 4.12**. Each type of ion has eight nearest neighbors which are the opposite type of ion, so the coordination number for each type of ion is 8. In addition, each unit cell is surrounded by six equivalent unit cells, each of which shares a face with its six neighbors. Since each of these unit cells contains a $Cs^+$ ion at its center, each $Cs^+$ ion has six next nearest neighbors that are $Cs^+$ ions, one in the center of each of the six neighbor unit cells.

**4.9    How many ions are in: (a) The CsCl unit cell?** The $Cs^+$ ion in the center of the unit cell, shown in **Figure 4.12**, is contained entirely within that cell. So, each unit cell contains one $Cs^+$ ion. The 1:1 stoichiometry of this compound demands that there must be exactly one $Cl^-$ contained in the cell. Recog-nizing that the $Cl^-$ ions at the corners of a unit cell contribute equally to eight different unit cells, each cell does indeed contain one $Cl^-$ ion ($8(1/8)$ $= 1$).

**(b) The sphalerite (ZnS) unit cell?** See the answer to **Example 4.1**.

**4.10    What structure is CsCl minus half the $Cs^+$ ions?** Several units cells of the CsCl structure are shown at the right. The shaded circles and the black circle are $Cl^-$ ions while the slightly smaller open circles are $Cs^+$ ions. Two of the four $Cs^+$ ions have been removed, following the instructions in the **Exercise**. If we focus our attention on the black $Cl^-$ ion, we see that it will have four nearest neighbor $Cs^+$ ions, the two shown and two more in a layer of unit cells that is adjacent to these four and out of the plane of the page. Those two $Cs^+$ ions will be contained in the lower left and upper right unit cells in the layer not shown, and the coordination geometry around the black $Cl^-$ ion (as well as *all* $Cl^-$ ions) will be tetrahedral. The structure we have formed is the fluorite structure, exhibited by $CaF_2$ and other ionic solids, and is shown in **Figure 4.14**.

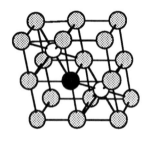

**4.11    Unit cell dimensions.** Refer to the drawings accompanying the answers to **End-of-Chapter Exercises 4.3** and **4.7**. Note that the unit cell edge length for the rock-salt structure is equal to the hypoteneuse of the isosceles right triangle drawn for the answer to **End-of-Chapter Exercise 4.3**. If the $Se^{2-}$ ions are just in contact in MgSe, then the legs of the isosceles right triangle are equal to twice the radius of a $Se^{2-}$ ion. Since the hypoteneuse (edge length) in MgSe is 5.45 Å, the ionic radius of $Se^{2-}$ is (5.45 Å)/(2√2) = 1.93Å. For the ionic solids CaSe, SrSe, and BaSe, the hypoteneuses (edge lengths) are 5.91 Å, 6.23 Å, and 6.62 Å, respectively, and are equal to twice the cation radius plus 5.45 Å, which is twice the radius of $Se^{2-}$. Thus, the ionic radii of $Ca^{2+}$, $Sr^{2+}$, and $Ba^{2+}$ are, respectively, (5.91 Å – 2(1.93) Å)/2 = 1.03 Å, (6.23 Å – 2(1.93) Å)/2 = 1.19 Å, and (6.62Å – 2(1.93) Å)/2 = 1.38 Å. This assumption of ion contact in certain limiting cases is one of the traditional ways of apportioning cation and anion radii in ionic compounds.

**4.12   Which are well described by an ionic model. (a) LiF.** Both elements in this compound are in Period 2, so the average principal quantum number, $n$, is 2. The difference in electronegativities, $\Delta\chi$, is 3.98 – 0.98 = 3.00 (data from **Table 1.9**). Therefore, LiF is way off the structure map in **Figure 4.21**, far to the right, and is certainly in the C.N. 6 region of the map. A very reasonable prediction is that LiF has the rock-salt structure. Since both ions are hard (nonpolarizable) and not too highly charged, an ionic model is a good one for this compound.

**(b)   RbBr.** The average $n$ is 4.5, and $\Delta\chi$ is 2.96 – 0.82 = 2.14. This puts RbBr in the C.N. 6 region of the map, so a reasonable prediction is that this compound has the rock-salt structure. An ionic model is still a reasonable one for RbBr, but not as good as it was for LiF (cf. **Table 4.4**).

**(c)   SrS.** The average $n$ is 4, and $\Delta\chi$ is 2.58 – 0.95 = 1.63. This puts SrS in the C.N. 6 region of the map, so a reasonable prediction is that this compound also has the rock-salt structure. Due to the polarizability of the $S^{2-}$ ion, an ionic model is probably only approximately valid for this compound. The lattice enthalpy calculated using an ionic model will probably be lower than the actual lattice enthalpy.

**(d)   BeO.** The average $n$ is 2, and $\Delta\chi$ is 3.44 – 1.57 = 1.87. This compound is right on the borderline between the C.N. 4 and C.N. 6 regions of the structure map. If we consider that *covalent* compounds containing a

Period 2 central atom are generally limited to four atoms around the small central atom, we predict that the two types of ions in BeO will exhibit C.N. 4. This is in fact the case: beryllium oxide has the wurtzite structure. An ionic model is not a good one for this compound (see the answer to **End-of-Chapter Exercise 4.6**.

**4.13   Coordination numbers of the ions in sphalerite and wurtzite.** Both of these structures are based on a close-packed array of the larger of the two ions ($S^{2-}$ in the case of sphalerite and wurtzite) with the smaller of the two ions ($Zn^{2+}$ in the case of sphalerite and wurtzite) in one-half of the tetrahedral holes. The sulfide ions are cubic close-packed in sphalerite and hexagonal close-packed in wurtzite. The coordination number is 4 for the cation and for the anion (see **Figures 4.13** and **4.15**).

**4.14   The stoichiometry of rutile.** The structure of a unit cell of the rutile form of $TiO_2$ is shown in **Figure 4.17**. The cell contains one $Ti^{4+}$ ion at the center, which belongs completely to the cell. It also contains 1/8 of the eight $Ti^{4+}$ ions at the corners of the unit cell. Thus, the unit cell contains two $Ti^{2+}$ ions. Four of the $O^{2-}$ ions surrounding the central $Ti^{4+}$ ion are on faces of the unit cell (although they are not at the centers of the faces). Two faces have two $O^{2-}$ ions each while the other four faces have none. One-half of these four oxide ions belong to the unit cell shown in the figure, while the other halves belong to adjacent cells. Since the other two $O^{2-}$ ions surrounding the central $Ti^{4+}$ ion are completely within the unit cell, the cell contains four net $O^{2-}$ ions $((4/2) + 2 = 4)$. Therefore, while the unit cell contains two $Ti^{4+}$ ions, it contains four $O^{2-}$ ions, and the 1:2 stoichiometry of rutile is consistent with the structure.

**4.15   The stoichiometry of perovskite.** The structure of a unit cell of perovskite is shown in **Figure 4.18**. The cell contains one $Ca^{2+}$ ion at its center. It also contains 1/8 of the eight $Ti^{4+}$ ions at the corners of the cell, for a total of 1 $Ti^{4+}$ ion and a Ca:Ti ratio of 1:1. Finally, the unit cell contains 1/4 of the twelve $O^{2-}$ ions that are centered on the twelve edges of the cubic unit cell, for a total of 3 $O^{2-}$ ions and a Ca:Ti:O ratio of 1:1:3. Therefore, the stoichiometry of perovskite, which is $CaTiO_3$, is consistent with the structure.


**4.16   Calculate $\Delta H_f$ for $KF_2$.** We can use a Born-Haber approach to this exercise (see the diagram accompanying the answer to **In-Chapter Exercise 4.3**). The unknown will be the enthalpy of formation, which is the bottom most arrow on the left side of the diagram. We will need to calculate or estimate $\Delta H_L$ for the hypothetical compound $KF_2$, which we can do using the Born-Mayer equation once we have estimated the radius of $K^{2+}$. To do this, we consult **Table 4.2**, and notice that $r(K^+) = 1.51$ Å, $r(Ca^{2+}) = 1.12$ Å, and $r(Sr^{2+}) = 1.25$ Å (these are C.N. 8 radii, since we were instructed to assume that $KF_2$ would have the fluorite structure). We can guess that $r(K^{2+}) > r(Ca^{2+})$, since $Z_{eff}(K^{2+}) < Z_{eff}(Ca^{2+})$. However, we can also guess that $r(K^{2+}) < r(Sr^{2+})$, since strontium has a higher principal quantum number than potassium. Thus, a reasonable estimate for $r(K^{2+})$ is 1.2 Å. Since the C.N. 4 radius of $F^-$ is 1.31 Å, the interionic distance in $KF_2$ is estimated to be 1.2 Å + 1.31 Å = 2.5 Å. Therefore:

$$\Delta H_L(KF_2) = (1.39 \text{ MJ mol}^{-1})(2/2.5 \text{ Å})(1 \text{ Å} -(0.345 \text{ Å})/(2.5 \text{ Å}))(2.519)$$

$$\Delta H_L(KF_2) = 2.4 \text{ MJ mol}^{-1}$$

(See **Example 4.5**; the Madelung constant for the $CaF_2$ structure is the last term in the expression for $\Delta H_L$, 2.519). This will be a favorable (negative) term as far as the formation of $KF_2$ (s) from K (s) and $F_2$ (g) is concerned. The only other favorable term is the electron affinity enthalpy change for two F atoms, $2(328 \text{ kJ mol}^{-1}) = 656 \text{ kJ mol}^{-1}$ (see **Table 1.8**). The positive enthalpy terms will be the sublimation enthalpy of potassium, 89 kJ mol$^{-1}$ (see **Figure 4.22**), the first *and* second ionization enthalpies of potassium, 419 kJ mol$^{-1}$ and 3069 kJ mol$^{-1}$ (see **Table 1.7**), respectively, and one-half the bond enthalpy of $F_2$, $(155 \text{ kJ mol}^{-1})/2 = 78 \text{ kJ mol}^{-1}$. Summing these together (see below) yields the estimated enthalpy of formation of $KF_2$, which is 599 kJ mol$^{-1}$, a rather large positive number. The predicted instability of $KF_2$ can be seen to be the result of the enormous second ionization enthalpy of potassium.

| Positive enthalpy terms in the formation of hypothetical $KF_2$ | | Negative enthalpy terms in the formation of hypothetical $KF_2$ | |
|---|---|---|---|
| Sub. of K | 89 kJ mol$^{-1}$ | Lattice enth. | $-2400$ kJ mol$^{-1}$ |
| Ionization of K | 3488 kJ mol$^{-1}$ | $e^-$ affin. of 2F | $-656$ kJ mol$^{-1}$ |
| 1/2 $F_2$ bond enth. | 78 kJ mol$^{-1}$ | | |
| Totals | 3655 kJ mol$^{-1}$ | | $-3056$ kJ mol$^{-1}$ |

$$\Delta H_f(KF_2) = 3655 \text{ kJ mol}^{-1} + (-3056 \text{ kJ mol}^{-1}) = 599 \text{ kJ mol}^{-1}$$

**4.17    Estimate the order of lattice energies for MgO, NaCl, and LiF.** Since the coulombic attraction accounts for the bulk of the lattice enthalpy, the compound with the largest value of $V_m$ (see **Section 4.7**) will have the largest lattice enthalpy. As long as all three compounds have the same structure (in this case they all have the rock–salt structure), $V_m \propto z_A z_B / d$. Without getting quantitative, we can see that the factor $z_A z_B$ is 4 for MgO but only 1 for NaCl and LiF. Therefore, the lattice enthalpy for MgO is the largest, because $d$ will hardly vary by a factor of 4 for the compounds under consideration. Between NaCl and LiF, the latter will have the larger lattice enthalpy, because $d$ will be smaller: Both $Li^+$ and $F^-$ are smaller than their counterparts in NaCl. Therefore, the order of increasing lattice enthalpies for these three ionic solids is NaCl < LiF < MgO.

**4.18    Which is more stable to thermal decomposition? (a) $MgCO_3$ or $CaCO_3$?** As discussed in the answer to **In-Chapter Exercise 4.7**, the enthalpy changes that are not constant for the two reactions

$$MgCO_3 \rightarrow MgO + CO_2 \qquad \text{and} \qquad CaCO_3 \rightarrow CaO + CO_2$$

are the four lattice enthalpies, for $MgCO_3$, MgO, $CaCO_3$, and CaO. Since $Mg^{2+}$ is smaller than $Ca^{2+}$, the enthalpy change $(\Delta H_L(MgCO_3) - \Delta H_L(MgO))$ is a *larger* negative number (and hence makes for a more favorable reaction) than the corresponding enthalpy change $(\Delta H_L(CaCO_3) - \Delta H_L(CaO))$.

Therefore, since the decomposition is more favorable for magnesium carbonate, calcium carbonate is more stable to thermal decomposition.

(b) $CsI_3$ or $N(CH_3)_4I_3$? The same reasoning may be applied here as above: the compound with the smaller cation will decompose more readily to a covalent compound and an ionic solid with a smaller anion (as large an anion as $I^-$ is, it is smaller than $I_3^-$). Thus, we need to compare the ionic radii of $Cs^+$ and $N(CH_3)_4^+$. The radius for $Cs^+$, taken from **Table 4.2**, is 1.67 Å (for C.N. 6). The radius for $N(CH_3)_4^+$ is not in any table in **Chapter 4**, but we can guess that it is larger than the radius for $Cs^+$. Therefore, since the decomposition is more favorable for cesium triiodide, tetramethyl-ammonium triiodide is more stable to thermal decomposition.

**4.19    Which is more soluble in water, LiCl or KCl?** As explained at the end of **Section 4.7**, salts of ions with widely different radii are generally soluble. Since the C.N. 6 radii of $Li^+$, $K^+$, and $Cl^-$ are 0.76 Å, 1.38 Å, and 1.81 Å, respectively (see **Table 4.2** and **Example 4.4**), LiCl is probably more soluble than KCl. In fact, 100 mL of water at 25 °C will dissolve approximately 70 g (1.7 mol) of LiCl but only about 30 g (0.40 mol) of KCl.

**4.20    What cation will result in the quantitative precipitation of $CO_3^{2-}$?** As in the **Exercise** above, a cation with a widely different radius than carbonate ion would lead to a soluble compound. Therefore, to achieve the lowest possible solubility, we would want a counterion of about the same size as carbonate ion. In addition, as explained at the end of **Section 4.7**, salts of the same charge type are less soluble than those of different charge types. So we will settle on a +2 cation of about the same size as $CO_3^{2-}$ ion. **Table 4.5** contains thermochemical radii of molecular ions, including that for $CO_3^{2-}$, which is 1.85 Å. Consulting **Table 4.2**, our best choice for a cation that will result in the lowest solubility carbonate salt is $Ba^{2+}$ (the ionic radius of $Ba^{2+}$ is not in the table, but it is the best choice since we can predict that it is larger than $Sr^{2+}$). The solubility of $BaCO_3$ is only 0.002 g (0.01 mmol) in 100 mL of water at 20 °C, whereas that of $MgCO_3$ is 0.01 g (0.12 mmol) in 100 mL of water at the same temperature.

**4.21**    **What anion will precipitate $[Co(NH_3)_5Cl]^{2+}$ the best, $SO_4^{2-}$ or $Cl^-$?** The least soluble salts are those in which the cation and anion have similar sizes and similar charges (see the end of **Section 4.7**). Therefore, sulfuric acid is a better choice than hydrochloric acid to cause the precipitation of this cobalt complex, because (i) sulfate is larger than chloride (2.30 Å vs. 1.81 Å, respectively; see **Example 4.4** and **Table 4.5**), and (ii) sulfate has the same charge type as the cobalt complex.

# CHAPTER 5

# BRØNSTED ACIDS AND BASES

| | | | | | | |
|---|---|---|---|---|---|---|
| Li | Be | B | C | N | | |
| Na | Mg | Al | Si | P | S | Cl |
| K | Ca | Ga | Ge | As | Se | Br |
| Rb | Sr | In | Sn | Sb | Te | I |
| Cs | Ba | Tl | Pb | Bi | | |

The *s*- and *p*-blocks of the Periodic Table, showing the elements that form basic oxides in plain type, those forming acidic oxides in outline type, and those forming amphoteric oxides in boldface type. Note the diagonal region from upper left to lower right that includes the elements forming amphoteric oxides. The elements Ge, Sn, Pb, As, Sb, and Bi form amphoteric oxides only in their lower oxidation state (II for Ge, Sn, and Pb, III for As, Sb, and Bi). They form acidic oxides in their higher oxidation state (IV for Ge, Sn, and Pb, V for As, Sb, and Bi).

## SOLUTIONS TO IN-CHAPTER EXERCISES

**5.1    Identifying acids and bases: (a) $HNO_3 + H_2O \rightarrow H_3O^+ + NO_3^-$.** The compound $HNO_3$ transfers a proton *to* water, so it is an acid. The nitrate ion is its conjugate base. In this reaction, $H_2O$ accepts a proton, so it is a base. The hydronium ion, $H_3O^+$, is its conjugate acid.

**(b) $CO_3^{2-} + H_2O \rightarrow HCO_3^- + OH^-$.** Carbonate ion accepts a proton from water, so it is a base. The hydrogen carbonate, or bicarbonate, ion is its conjugate acid. In this reaction, $H_2O$ donates a proton, so it is an acid. Hydroxide ion is its conjugate base.

(c) $NH_3 + H_2S \rightarrow NH_4^+ + HS^-$. Ammonia accepts a proton from hydrogen sulfide, so it is a base. The ammonium ion, $NH_4^+$, is its conjugate acid. Since hydrogen sulfide donated a proton, it is an acid, while $HS^-$ is its conjugate base.

**5.2     Arrange in order of increasing acidity:** $[Na(H_2O)_6]^+$, $[Sc(H_2O)_6]^{3+}$, $[Mn(H_2O)_6]^{2+}$, **and** $[Ni(H_2O)_6]^{2+}$. Since the acid strength of aqua acids increases as the electrostatic parameter, $\xi = z/(r + d)$, the strongest acid will have the highest charge, at least for a group of acids for which $r + d$ does not vary too much. Thus $[Na(H_2O)_6]^+$, with the lowest charge, will be the weakest of the four aqua acids, and $[Sc(H_2O)_6]^{3+}$, with the highest charge, will be the strongest. The remaining two aqua acids have the same charge, and so the one with the smaller ionic radius, $r$, will have the smaller value of $r + d$ and hence the greater acidity. Since $Ni^{2+}$ has a greater $Z_{eff}$ than $Mn^{2+}$, it has a smaller radius, and so $[Ni(H_2O)_6]^{2+}$ is more acidic than $[Mn(H_2O)_6]^{2+}$. The order of increasing acidity is $[Na(H_2O)_6]^+ < [Mn(H_2O)_6]^{2+} < [Ni(H_2O)_6]^{2+} < [Sc(H_2O)_6]^{3+}$.

**5.3     Predict $pK_a$ values: (a) $H_3PO_4$.** Pauling's first rule for predicting the $pK_a$ of a mononuclear oxoacid is $pK_a \approx 8 - 5p$ (where $p$ is the number of oxo groups attached to the central element).

Since $p = 1$, the predicted value of $pK_a$ for $H_3PO_4$ is $8 - 5 \times 1 = 3$. The actual value, given in **Table 5.4**, is 2.1.

**(b) $H_2PO_4^-$.** Pauling's second rule for predicting the $pK_a$ of a mononuclear oxoacid is that successive $pK_a$ values for polyprotic acids increase by five units for each successive proton transfer. Since $pK_a(1)$ for $H_3PO_4$ was predicted to be 3 (see above), the predicted value of $pK_a$ for $H_2PO_4^-$, which is $pK_a(2)$ for $H_3PO_4$, is $3 + 5 = 8$. The actual value, given in **Table 5.4**, is 7.4.

(c) **$HPO_4^{2-}$.** The $pK_a$ for $HPO_4^{2-}$ is the same as $pK_a(3)$ for $H_3PO_4$, so the predicted value is $3 + 2 \times 5 = 13$. The actual value, given in **Table 5.4**, is 12.3.

**5.4 What happens to Ti(IV) in aqueous solution as the pH is raised?** According to **Figure 5.6**, Ti(IV) is amphoteric. Treatment of an aqueous solution containing Ti(IV) ions with ammonia causes the precipitation of $TiO_2$, but further treatment with NaOH causes the $TiO_2$ to redissolve.

## SOLUTIONS TO END-OF-CHAPTER EXERCISES

**5.1 Sketch an outline of the $s$ and $p$ blocks of the Periodic Table, showing the elements that form acidic, basic, and amphoteric oxides.** See the diagram at the beginning of this chapter. If you cannot write out the $s$- and $p$-blocks from memory, you should spend some time learning that part of the periodic table. This knowledge will permit you to integrate many chemical facts into a logical pattern of trends.

**5.2 Identify the conjugate bases of the following acids: (a) $[Co(NH_3)_5(OH_2)]^{3+}$.** A conjugate base is a species with one fewer proton than the parent acid. Therefore, the conjugate base in this case is $[Co(NH_3)_5(OH)]^{2+}$, shown below (L = $NH_3$).

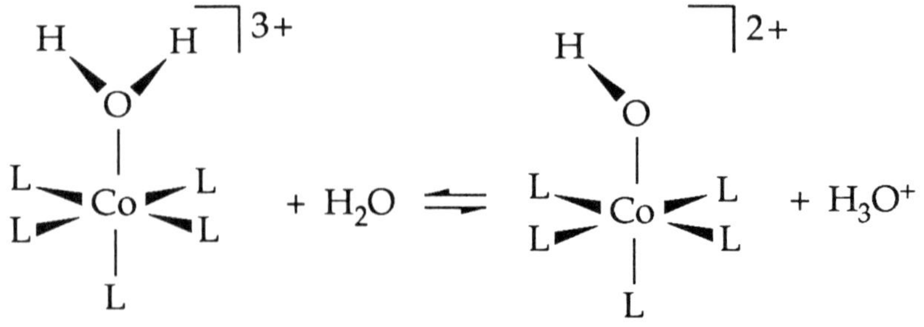

$$\left[\begin{array}{c} H \diagdown \diagup H \\ O \\ | \\ L \diagup Co \diagdown L \\ L \diagup | \diagdown L \\ L \end{array}\right]^{3+} + H_2O \rightleftharpoons \left[\begin{array}{c} H \diagdown \\ O \\ | \\ L \diagup Co \diagdown L \\ L \diagup | \diagdown L \\ L \end{array}\right]^{2+} + H_3O^+$$

**(b) $HSO_4^-$.** The conjugate base is $SO_4^{2-}$.

**(c) $CH_3OH$.** The conjugate base is $CH_3O^-$.

**(d) $H_2PO_4^-$.** The conjugate base is $HPO_4^{2-}$.

**(e) $Si(OH)_4$.** The conjugate base is $SiO(OH)_3^-$.

**(f) $HS^-$.** The conjugate base is $S^{2-}$.


**5.3    Identify the conjugate acids of the following bases: (a) $C_5H_5N$ (pyridine).**    A conjugate acid is a species with one more proton than the parent base.  Therefore, the conjugate acid in this case is the pyridinium ion, $C_5H_6N^+$, shown below.

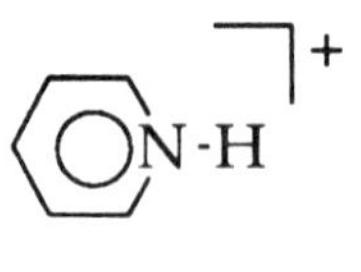

$C_5H_6N^+$ 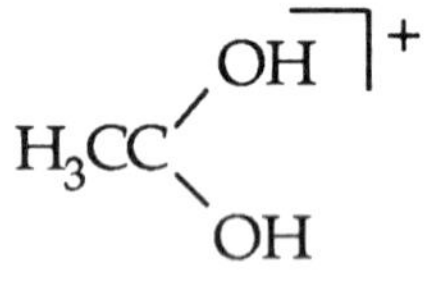                          $CH_3C(OH)_2^+$

**(b) $HPO_4^{2-}$.** The conjugate acid is $H_2PO_4^-$.

**(c) $O^{2-}$.** The conjugate acid is $OH^-$.

**(d) $CH_3COOH$.** The conjugate acid is $CH_3C(OH)_2^+$, shown above.

**(e) [Co(CO)₄]⁻.** The conjugate acid is $HCo(CO)_4$, shown below.

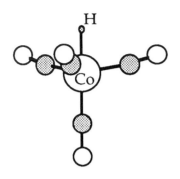

A drawing of the $HCo(CO)_4$ molecule, the conjugate acid of the tetrahedral $Co(CO)_4^-$ anion. The C atoms of the CO ligands are bound to the Co atom. The O atoms are unshaded.

**(f) CN⁻.** The conjugate acid is HCN.

**5.4    List the bases HS⁻, F⁻, I⁻, and NH₂⁻ in order of increasing proton affinity.** We make use of **Tables** 5.2 and 5.3 to answer this question. The species with the greatest proton affinity will be the strongest base, and its conjugate acid will be the weakest acid. The weakest acid will have the smallest value of $K_a$ (or the most positive value of $pK_a$). Since **Table 5.2** shows that HI is a stronger acid than HF which is a stronger acid than $H_2S$, a partial order of proton affinity is I⁻ < F⁻ < HS⁻. **Table 5.3** lists values of gas–phase and solution proton affinities. For either situation, the proton affinity of NH⁻ is greater than that of F⁻, which in turn is greater than that of I⁻. Therefore, our final list, in order of increasing proton affinity, is I⁻ < F⁻ < HS⁻ <NH₂⁻.

**5.5    List the acids H₂O, HCl, HI, and CH₄ in order of increasing gas–phase acidity.** The most acidic molecule will have the conjugate base with the smallest gas-phase proton affinity, and the least acidic molecule will have the conjugate base with the greatest gas-phase proton affinity. Reference to **Table 5.3** shows that the proton affinity of $CH_3^-$ (the conjugate base of $CH_4$) is greater than that of OH⁻, which is greater than that of F⁻. The gas–phase proton affinity of I⁻ is the smallest for the four anions in question. Thus, a list of the four acids, in order of increasing gas-phase acidity, is $CH_4 < H_2O < HF < HI$.

**5.6    Calculating proton transfer enthalpy changes.** The net reaction can be broken down into two single proton transfer steps, the enthalpy changes for which can be found in **Table 5.3**:

$NH_4^+$ (g) $\rightarrow$ $NH_3$ (g) + $H^+$ (g)      $\Delta H = -\Delta H_p(NH_3) = A_p(NH_3) = 865$ kJ mol$^{-1}$

$Cl^-$ (g) + $H^+$ (g) $\rightarrow$ HCl (g)        $\Delta H = \Delta H_p(Cl^-) = -A_p(Cl^-) = -1393$ kJ mol$^{-1}$

---

$NH_4^+$ (g) + $Cl^-$ (g) $\rightarrow$ $NH_3$ (g) + HCl (g)              $\Delta H_{rxn} = -528$ kJ mol$^{-1}$

This very favorable enthalpy change is surprising, given that HCl is a *strong* acid. Of course, it is hydrochloric acid, an aqueous solution of hydrogen chloride (HCl), that is a strong acid. If this reaction is carried out in aqueous solution instead of in the gas phase, the enthalpy change will be large *and positive*. Solvation effects (*hydration* effects in the case of water as the solvent) have been ignored in our calculation of the net enthalpy change, $\Delta H_{rxn}$. The reactants, which are ions, will be strongly stabilized in water relative to the products, which are neutral molecules, and the net enthalpy change will switch from negative (gas-phase) to positive (aqueous solution).

**5.7    Which bases are too strong or too weak to be studied experimentally?** **(a)  $CO_3^{2-}$ $O^{2-}$, $ClO_4^-$, and $NO_3^-$ in water.** We can interpret the term "studied experimentally" to mean that the base in question exists in water (i.e. it is not completely protonated to its conjugate acid) *and* that the base in question can be partially protonated (i.e. it is not so weak that the strongest acid possible in water, $H_3O^+$, will fail to produce a measurable amount of the conjugate acid). Using these criteria, the base $CO_3^{2-}$ is of directly measurable base strength, since the equilibrium $CO_3^{2-}$ + $H_2O$ $\rightleftharpoons$ $HCO_3^-$ + $OH^-$ produces measurable amounts of reactants and products. The base $O^{2-}$, on the other hand, is completely protonated in water to produce $OH^-$, so the oxide ion is too strong to be studied experimentally in water. The bases $ClO_4^-$ and $NO_3^-$ are conjugate bases of very strong acids, which are completely deprotonated in water. Therefore, since it is not possible to protonate either perchlorate or nitrate ion in water, they are too weak to be studied experimentally.

(b) $HSO_4^-$, $NO_3^-$, and $ClO_4^-$, in $H_2SO_4$. The hydrogen sulfate ion, $HSO_4^-$, is the strongest base possible in liquid sulfuric acid. However, since acids can protonate it, it is not too strong to be studied experimentally. Nitrate ion is a weaker base than $HSO_4^-$, a consequence of the fact that its conjugate acid, $HNO_3$, is a stronger acid than $H_2SO_4$. However, nitrate is not so weak that it cannot be protonated in sulfuric acid, so $NO_3^-$ is of directly measurable base strength in liquid $H_2SO_4$. On the other hand, $ClO_4^-$, the conjugate base of one of the strongest known acids, is so weak that it cannot be protonated in sulfuric acid, and hence cannot be studied in sulfuric acid.

**5.8    Is the –CN group electron donating or withdrawing?** A comparison of the aqueous $pK_a$ values is necessary to answer this question. These are:

HOCN, 4      $H_2NCN$, 10.5        $CH_3CN$, 20

$H_2O$, 14      $NH_3$, ?      $CH_4$, ?

Judging from the comparison of HOCN and $H_2O$, we conclude that the –CN group is electron withdrawing, since it stabilizes the $OCN^-$ anion relative to the $OH^-$ anion and makes HOCN a much stronger acid (lower $pK_a$) than $H_2O$. To see if this trend continues for the other two compounds, we must estimate $pK_a$ for $NH_3$ and $CH_4$. We can do this by using the solution proton affinity data ($A_p'$ values) in **Table 5.3**. We know that the enthalpy change for the reaction is the difference between $A_p'(H_2O)$ and $A_p'(OH^-)$:

$$2\,H_2O \rightleftharpoons H_3O^+ + OH^- \qquad \Delta H = \Delta A_p' = 1188 - 1130 \text{ kJ mol}^{-1} = 58 \text{ kJ mol}^{-1}$$

Ammonia is a less acidic substance than $H_2O$ in aqueous solution, and to estimate how much less we use the equation:

$$H_2O + NH_3 \rightleftharpoons H_3O^+ + NH_2^- \qquad \begin{aligned} \Delta H &= \Delta A_p' = 1351 - 1130 \text{ kJ mol}^{-1} \\ \Delta H &= 221 \text{ kJ mol}^{-1} \end{aligned}$$

If we call the equilibrium constant for the first equation $K_1$ and that for the second equation $K_2$, we can use the following expression, derived from the material in the **Further Information** section of this chapter, assuming that $\Delta\Delta H = \Delta\Delta G$:

$$K_1/K_2 = \exp(-\Delta\Delta G/RT) = \exp(-\Delta\Delta H/RT)$$

$$K_1/K_2 = \exp(221 - 168 \text{ kJ mol}^{-1})/(8.314 \text{ J K}^{-1} \text{ mol}^{-1})(298 \text{ K}) = 3.8 \times 10^{-30}$$

From this value of $K_1/K_2$ we can calculate $\Delta pK_a$, which is $-\log(3.8 \times 10^{-30})$ or 29.41. Since $pK_a$ for $H_2O$ (= $pK_w$) is 14, then $pK_a$ for $NH_3$ is ~29 + 14 = 41. This is considerably greater than the $pK_a$ for $H_2NCN$, so once again the cyano-containing compound is a stronger acid and we conclude that the $-CN$ group is an electron withdrawing group. Furthermore, since $A_p'$ for $CH_3^-$ is greater than $A_p'$ for $NH_2^-$, this conclusion will be borne out for $CH_4$ vs. $CH_3CN$ as well.

**5.9    Is the $pK_a$ for $HAsO_4^{2-}$ consistent with Pauling's rules?** Pauling's first rule for predicting the $pK_a$ of a mononuclear oxoacid is $pK_a \approx 8 - 5p$ (where $p$ is the number of oxo groups attached to the central element).

Since $p = 1$, the predicted value of $pK_a(1)$ for $H_3PO_4$ is $8 - 5 \times 1 = 3$.

Pauling's second rule for predicting the $pK_a$ of a mononuclear oxoacid is that successive $pK_a$ values for polyprotic acids increase by five units for each successive proton transfer. Since $pK_a(1)$ for $H_3AsO_4$ was predicted to be 3, the predicted value of $pK_a$ for $HAsO_4^{2-}$, which is $pK_a(3)$ for $H_3AsO_4$, is $3 + 2 \times 5 = 13$. The actual value, which differs by 1.5 $pK_a$ units, is 11.5. This illustrates that Pauling's rules are only approximate.

**5.10    Account for the trends in the $pK_a$ values of the conjugate acids of $SiO_4^{4-}$, $PO_4^{3-}$, $SiO_4^{2-}$, and $ClO_4^-$.** The structures of these four anions, which can be determined to be tetrahedral using VSEPR, are shown below. As can be seen, the charge on the anions decreases from $-4$ for the silicon-containing species to $-1$ for the chlorine-containing species. The charge differences alone would make $SiO_4^{4-}$ the most basic species. Hence $HSiO_4^{3-}$ is the least acidic conjugate acid. The acidity of the four conjugate acids increases in the order $HSiO_4^{3-} < HPO_4^{2-} < HSO_4^- < HClO_4$.

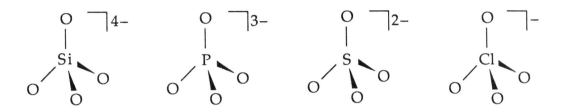

**5.11   Which of the following is the stronger acid?   (a)   $[Fe(OH_2)_6]^{3+}$ or $[Fe(OH_2)_6]^{2+}$?** The Fe(III) complex, $[Fe(OH_2)_6]^{3+}$, is the stronger acid by virtue of the higher charge. The electrostatic parameter, $\xi = z^2/(r + d)$, will be considerably higher for $z = 3$ than for $z = 2$. The minor decrease in $r + d$ on going from the Fe(II) to the Fe(III) species will enhance the differences in $\xi$ for the two species.

**(b)   $[Al(OH_2)_6]^{3+}$ or $[Ga(OH_2)_6]^{3+}$?** In this case, $z$ is the same but $r + d$ is different. Since the ionic radius, $r$, is smaller for Period 3 $Al^{3+}$ than for Period 4 $Ga^{3+}$, $r + d$ for $[Al(OH_2)_6]^{3+}$ is smaller than $r + d$ for $[Ga(OH_2)_6]^{3+}$ and the aluminum–containing species is more acidic.

**(c)   $Si(OH)_4$ or $Ge(OH)_4$?** As in part **(b)**, above, $z$ is the same but the $r + d$ parameter is different for these two compounds. The comparison here is also between species containing Period 3 and Period 4 central atoms in the same Group, and the species containing the smaller central atom, $Si(OH)_4$, is more acidic.

**(d)   $HClO_3$ or $HClO_4$?** These two acids are shown below. According to Pauling's rule 1 for mononuclear oxoacids, the species with more oxo groups has the lower $pK_a$ and is the stronger acid. Thus, $HClO_4$ is a stronger acid than $HClO_3$. Note that the oxidation state of the central chlorine atom in the stronger acid (VII) is higher than in the weaker acid (V).

$$HO-Cl(=O)(=O)$$

HClO$_3$                          HClO$_4$

**(e) H$_2$CrO$_4$ or HMnO$_4$?** As in part **(d)**, above, the oxidation states of these two acids are different, VI for the chromium atom in H$_2$CrO$_4$ and VII for the manganese atom in HMnO$_4$. The species with the higher central-atom oxidation state, HMnO$_4$, is the stronger acid. Note that this acid has more oxo groups, three, than H$_2$CrO$_4$, which has two.

**(f) H$_3$PO$_4$ or H$_2$SO$_4$.** The oxidation state of sulfur in H$_2$SO$_4$ is VI while the oxidation state of phosphorus in H$_3$PO$_4$ is only V. Furthermore, sulfuric acid has two oxo groups attached to the central sulfur atom while phosphoric acid has only one oxo group attached to the central phosphorus atom. Therefore, on both counts (which by now we can see are really manifestations of the same thing) H$_2$SO$_4$ is a stronger acid than H$_3$PO$_4$.

**5.12    Arrange the following oxides in order of increasing basicity: Al$_2$O$_3$, B$_2$O$_3$, BaO, CO$_2$, Cl$_2$O$_7$, and SO$_3$.** First we pick out the intrinsically acidic oxides, since these will be the *least* basic. The compounds B$_2$O$_3$, CO$_2$, Cl$_2$O$_7$, and SO$_3$ are acidic, since the central element for each of them is found in the acidic region of the periodic table (see the *s* and *p* blocks diagram at the beginning of this chapter as well as **Figure 5.5**). The most acidic compound, Cl$_2$O$_7$, has the highest central–atom oxidation state, VII, while the least acidic, B$_2$O$_3$, has the lowest, III. Of the remaining compounds, Al$_2$O$_3$ is amphoteric, which puts it on the borderline between acidic and basic oxides, and BaO is basic. Therefore, a list of these compounds in order of increasing basicity is Cl$_2$O$_7$ < SO$_3$ < CO$_2$ < B$_2$O$_3$ < Al$_2$O$_3$ < BaO.

**5.13    Arrange the following in order of increasing acidity: HSO$_4^-$, H$_3$O$^+$, H$_4$SiO$_4$, CH$_3$GeH$_3$, NH$_3$, and HSO$_3$F.** The weakest acids, CH$_3$GeH$_3$ and NH$_3$, are easy to pick out of this group since they do not contain any –OH bonds. Ammonia is the weaker acid of the two, since it has a lower central-atom oxidation state, III, than that for the germanium atom in $CH_3GeH_3$, which is IV. Of the remaining species, note that $HSO_3F$ is very similar to $H_2SO_4$ as far as structure and sulfur oxidation state (VI) are concerned, so it is reasonable to suppose that $HSO_3F$ is a very strong acid, which it is. The anion $HSO_4^-$ is a considerably weaker acid than $HSO_3F$, for the same reason that it is a considerably weaker acid than $H_2SO_4$, namely Pauling's rule 2 for mononuclear oxoacids. Since $HSO_4^-$ is not completely deprotonated in water, it is a weaker acid than $H_3O^+$, which is the strongest possible acidic species in water. Finally, it is difficult to place exactly $Si(OH)_4$ in this group. It is certainly more acidic than $NH_3$ and $CH_3GeH_3$, and it turns out to be *less* acidic than $HSO_4^-$, despite the negative charge of the latter species. Therefore, a list of these species in order of increasing acidity is $NH_3 < CH_3GeH_3 < H_4SiO_4 < HSO_4^- < H_3O^+ < HSO_3F$.

The structures of $H_2SO_4$ and $HSO_3F$.

**5.14 Which aqua ion is the stronger acid, $Na^+$ or $Ag^+$?** Even though these two ions have about the same ionic radius, silver(I)–$OH_2$ bonds are much more covalent than sodium(I)–$OH_2$ bonds, a common feature of the chemistry of *d*-block vs. *s*-block metal ions. The greater covalence of the silver(I)–$OH_2$ bonds has the effect of delocalizing the positive charge of the cation over the whole aqua complex. As a consequence, the departing proton is repelled more by the positive charge of $Ag^+$ (*aq*) than by the positive charge of $Na^+$ (*aq*), and the former ion is the stronger.

**5.15 Which of the following elements form oxide polyanions or polycations: Al, As, Cu, Mo, Si, B, Ti?** As discussed in **Sections 5.8, 5.9,** and **5.10,** the aqua ions of metals that have basic or amphoteric oxides generally undergo polymerization to polycations. The elements Al, Cu, and Ti fall into this category. On the other hand, polyoxoanions (oxide polyanions) are important for some of the early *d*-block metals, especially for V, Mo, and W in high oxidation states. Furthermore, many of the *p*-block elements form polyoxoanions, including As, B, and Si.

**5.16   The change in charge upon aqua ion polymerization.** Examples of aqua ion polymerization is:

$$2\,[Al(OH_2)_6]^{3+} + H_2O \rightarrow [(H_2O)_5Al\text{–}O\text{–}Al(OH_2)_5]^{4+} + H_3O^+$$

The charge per aluminum atom is +3 for the mononuclear species on the left hand side of the equation but only +2 for the dinuclear species on the right hand side. Thus, poly*cation* formation reduces the average positive charge per central M atom by +1 per M.

$$
\begin{array}{c}
\hspace{6cm}\rceil\,4+ \\[4pt]
H_2O \;\; OH_2 \qquad H_2O \;\; OH_2 \\[-2pt]
\;\;\;\diagdown\!\diagup \qquad\qquad \diagdown\!\diagup \\[-4pt]
H_2O - Al - O - Al - OH_2 \\[-2pt]
\;\;\diagup\,| \qquad\qquad \diagup\,| \\[-2pt]
H_2O \;\; OH_2 \quad H_2O \;\; OH_2
\end{array}
$$

**5.17   Write balanced equations for the formation of $P_4O_{12}^{4-}$ from $PO_4^{3-}$ and for the formation of $[(H_2O)_4Fe(OH)_2Fe(OH_2)_4]^{4+}$ from $[Fe(OH_2)_6]^{3+}$.** The two balanced equations are shown below. Note that the condensation reactions involve a neutralization of charge, either by adding $H^+$ to a highly charged anion or by removing $H^+$ from a highly charged cation. The structure of $P_4O_{12}^{4-}$, which is called *cycle*tetrametaphosphate, is also shown below.

$$4PO_4^{3-} + 8H_3O^+ \rightarrow P_4O_{12}^{4-} + 12H_2O$$

$$2[Fe(OH_2)_6]^{3+} \rightarrow [(H_2O)_4Fe(OH)_2Fe(OH_2)_4]^{4+}\; 2H_3O^+$$

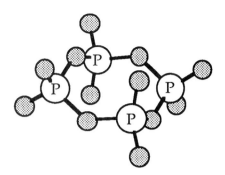

The structure of the $[P_4O_{12}]^{4-}$ ion in the salt $[NH_4]_2[P_4O_{12}]$.

**5.18 More Balanced equations: (a) $H_3PO_4$ and $Na_2HPO_4$.** You can use the successive $K_a$ values for phosphoric acid to estimate the equilibrium constant for the equilibrium below:

$$H_3PO_4 + HPO_4^{2-} \rightleftharpoons 2H_2PO_4^- \qquad K = ?$$

The three $pK_a$ values are found in **Table 5.4** and are 2.1, 7.4, and 12.3, so the three $K_a$ values are $8 \times 10^{-3}$ $(K_{a1})$, $4 \times 10^{-8}$ $(K_{a2})$, and $5 \times 10^{-13}$ $(K_{a3})$. The equilibrium above is the sum of the two equilibria below, so $K$ for the equilibrium above is the product $(K_{a1})(1/K_{a2}) = 2 \times 10^5$.

$$H_3PO_4 + H_2O \rightleftharpoons H_2PO_4^- + H_3O^+ \qquad K_{a1} = 8 \times 10^{-3}$$

$$HPO_4^{2-} + H_3O^+ \rightleftharpoons H_2PO_4^- + H_2O \qquad 1/K_{a2} = 3 \times 10^7$$

**(b) $CO_2$ and $CaCO_3$.** Successive $K_a$ values can also be used to show that the equilibrium below lies to the right.

$$CO_2 + CaCO_3 + H_2O \rightleftharpoons Ca^{2+} + 2HCO_3^-$$

**5.19 Identify the point groups of $[W_6O_{19}]^{6-}$ and $[Fe_2(OH_2)_8(OH)_2]^{4+}$.** These complexes are shown in **Structural Drawings 14** and **11**, respectively. The tungsten complex possesses three $C_4$ axes, four $C_3$ axes, and a center of symmetry, as well as many other symmetry elements. However, the presence of these three classes of symmetry elements are sufficient to uniquely define the point group of this anion as $O_h$ (see the decision tree in

Figure 2.12).    The iron complex possesses exactly three mutually perpendicular $C_2$ axes and a center of symmetry, and these two classes alone are sufficient to uniquely define the point group of this cation as $D_{2h}$.

# CHAPTER 6

# LEWIS ACIDS AND BASES

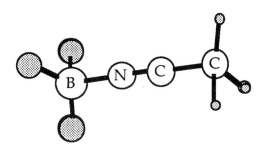

A simple Lewis acid-base adduct, formed by mixing the Lewis acid $BF_3$ with the Lewis base MeCN. In this adduct, the structure of the Lewis acid has changed from planar to pyramidal, while the structure of the Lewis base has hardly changed at all.

## SOLUTIONS TO IN-CHAPTER EXERCISES

**6.1    Identify the acids and bases.**  A general rule that works in many, but not all, instances is that negatively charged ions are Lewis bases and positively charged ions are Lewis acids.  This rule certainly works in the three parts to this exercise.  However, be alert for the possibility that a charged species may be neither acidic nor basic, just as an electrically neutral species may be neither acidic nor basic.

(a)  $FeCl_3 + Cl^- \rightarrow [FeCl_4]^-$.  The acid $FeCl_3$ forms a complex, $[FeCl_4]^-$, with the base $Cl^-$.

(b)  $I^- + I_2 \rightarrow I_3^-$.  The acid $I_2$ forms a complex, $I_3^-$, with the base $I^-$.

(c) $[SnCl_3]^- + (CO)_5MnCl \rightarrow (CO)_5Mn-SnCl_3 + Cl^-$. The acid, $Mn(CO)_5^+$, is displaced from its complex with the base $Cl^-$ by the base $[:SnCl_3]^-$, and the new complex $(CO)_5Mn-SnCl_3$ is formed.

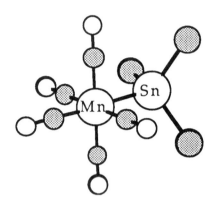

A Drawing of the $(CO)_5Mn-SnCl_3$ molecule, a complex of $[Mn(CO)_5]^+$ and $[SnCl_3]^-$. The carbon atoms of the CO ligands are bound to the Mn atom. The O atoms are unshaded.

In part (c), both the Lewis bases are species that are stable whether or not they are part of the acid-base complex. This is true for many, but not all, Lewis acids. In this case, the Lewis acid is not a stable species that has an independent existence. That is, the cation $[Mn(CO)_5]^+$ cannot be isolated as a simple salt, only as complexes. A more familiar example of a Lewis acid that does not have an independent existence in solution or in the solid state is the proton, or hydrogen ion, $H^+$. Recall that the free proton is always bound to a solvent molecule or another basic species (**Section 5.1**).

**6.2    Aluminosilicate minerals: Rb, Cr, and Sr.    Sulfides: Cd, Pb, and Pd.** The bases in aluminosilicate minerals are hard silicate oxo anions, while the base in sulfide minerals is the soft base $S^{2-}$. Given these two choices, hard metals will be found in aluminosilicate minerals and soft metals will be found in sulfides. Of the list of metals given, Rb, Cr, and Sr are hard, and are found in aluminosilicates. The metals Rb and Sr are alkali metals, all of which are hard. Chromium is an early *d*-block metal, and all of these are hard too. The metals Cd, Pb, and Pd are soft, and are found in sulfides. Lead is a heavy *p*-block metal, all of which are soft. The metals Cd and Pd are 2nd row late *d*-block metals. Along with the period 6 *d*-block metals, all of these are soft too.

**6.3**     **(a) $C_2H_2$ from $CaC_2$.** The reaction $CaC_2 + H_2O \rightarrow CaO + C_2H_2$ could be used to prepare ethyne (acetylene) from calcium carbide. Both $CaC_2$ and CaO are ionic compounds. The hard acid $Ca^{2+}$ prefers the hard base $O^{2-}$ to the soft carbon base $C_2^{2-}$. While the proton, $H^+$, is also a hard acid, it can form strong covalent bonds with both bases. Another factor is that –OH compounds are more acidic than –CH compounds.

   **(b) $P(CH_3)_3$ from $PCl_3$.** The reaction $PCl_3 + 3CH_3Li \rightarrow P(CH_3)_3 +$ $3LiCl$ could be used to prepare trimethylphosphine. The products of this reaction are a molecular substance containing a soft acid and a soft base ($P(CH_3)_3$) and an ionic substance containing a hard acid and a relatively hard base (LiCl).

**6.4**     **Identifying steric effects.** The two Lewis bases 2-Me-THF and 3-Me-THF are electronically very similar, since both are tetrahydrofurans with a single methyl substituent. However, in 2-Me-THF, the methyl group is on the carbon atom next to the oxygen atom (that is, the methyl group is $\alpha$ to the oxygen atom), and some steric crowding between the methyl group and the acid, $BF_3$, weakens the B–O bond in the resulting adduct relative to the B–O bond in the adduct of $BF_3$ and 3-Me-THF, where steric crowding due to the methyl group is not a factor.

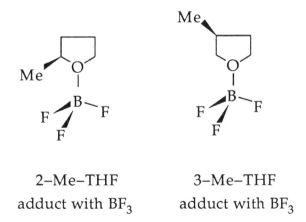

2–Me–THF           3–Me–THF
adduct with $BF_3$      adduct with $BF_3$

**6.5**     **Select the better solvent. (a)** $[Co(NH_3)_5Cl]^{2+} + S \rightarrow [Co(NH_3)_5S]^{3+} +$ $Cl^-$, S = MeOH or $H_2O$? The solvent for this reaction must act as a donor toward $Co^{3+}$ and as a dielectric medium to promote the separation of the cationic cobalt complex from $Cl^-$. Water and MeOH have similar donor numbers, but $H_2O$ has a much higher relative permittivity, $\varepsilon_r$ (Table 6.2). Therefore, $H_2O$ is the better solvent in this case.

(b) $AgClO_4(s) \rightarrow Ag^+(S) + ClO_4^-(S)$, S = benzene or cyclohexane? Even though benzene is a poor donor and has a low relative permittivity, cyclohexane is a poorer donor and has a lower $\varepsilon_r$. Therefore, benzene is the better solvent for dissolving silver(I) perchlorate. The complex of $Ag^+$ and a molecule of toluene is shown below. Aromatic hydrocarbons use their $\pi$ electrons to form $\sigma$ bonds with Lewis acids.

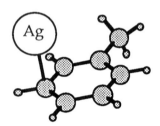

The $Ag^+$ ion is above the plane of the toluene molecule and is bonded more strongly to one carbon atom than to the others. In toluene solution, $Ag^+$ would be bonded to more than one toluene molecule.

**6.6    Using the Drago-Wayland equation.** The enthalpy of complex (adduct) formation, $\Delta H$, can be calculated using the Drago-Wayland equation and the acid/base parameters in **Table 6.3**:

$$\Delta H = -[E_A E_B + C_A C_B]$$

(a) ICl–(*p*-dioxane). $\Delta H = -[(10.43)(2.23) + (1.70)(4.87)] = -31.5$ kJ $mol^{-1}$ (the experimental value is –31.4 kJ $mol^{-1}$).

(b) $Me_3B$–$NH_2Me$. $\Delta H = -[(12.55)(2.66) + (3.48)(12.03)] = -75.2$ kJ $mol^{-1}$ (the experimental value is –73.7 kJ $mol^{-1}$).

## SOLUTIONS TO END-OF-CHAPTER EXERCISES

**6.1    Identifying elements that form Lewis acids in lower oxidation states.**

| B | C | N | O | F |
|---|---|---|---|---|
| Al | Si | P | S | Cl |
| Ga | Ge | As | Se | Br |
| In | Sn | Sb | Te | I |
| Tl | Pb | Bi | | |

Most of the *p*-block elements form Lewis acids in one of their lower oxidation states, except for the lightest members and some of the next lightest members of the these groups. Examples are as follows: Ga, In, and Tl all form +1 cations; the dichlorides of Ge, Sn, and Pb are Lewis acids; the trifluorides of P, As, Sb, and Bi are Lewis acids ($PF_3$ and $AsF_3$ are also Lewis bases towards *d*-block metals, so these compounds are amphoteric); the dioxides of S, Se, and Te are Lewis acids; the trifluorides of Cl, Br, and I are Lewis acids.

**6.2    Identifying acids and bases. (a) $SO_3 + H_2O \rightarrow HSO_4^- + H^+$.** The acids in this reaction are the Lewis acids $SO_3$ and $H^+$ and the base is the Lewis base $OH^-$. The complex (or adduct) $HSO_4^-$ is formed by the displacement of the proton from the hydroxide ion by the stronger acid $SO_3$. In this way, the water molecule is thought of as an adduct of $H^+$ and $OH^-$. Since the proton must be bound to a solvent molecule, even though this fact is not explicitly shown in the reaction, the water molecule exhibits Brønsted acidity. Note that it is easy to tell that this is a displacement reaction instead of just a complex formation reaction because, while there is only one base in the reaction, there are *two* acids. A complex formation reaction only occurs with a single acid and a single base. A double displacement, or metathesis, reaction only occurs with two acids and two bases.

**(b) $Me[B_{12}]^- + Hg^{2+} \rightarrow [B_{12}] + MeHg^+$** ([B12] designates the Co center of the macrocyclic complex called vitamin B12). This is a displacement reaction. The Lewis acid $Hg^{2+}$ displaces the Lewis acid $[B_{12}]$ from the Lewis base $CH_3^-$.

(c) $KCl + SnCl_2 \rightarrow K^+ + [SnCl_3]^-$. This is also a displacement reaction. The Lewis acid $SnCl_2$ displaces the Lewis acid $K^+$ from the Lewis base $Cl^-$.

(d) $AsF_3$ (g) + $SbF_5$ (g) $\rightarrow$ $[AsF_2][SbF_6]$. Even though this reaction is the formation of an ionic substance, it is *not* simply a complex formation reaction. It is a displacement reaction. The very strong Lewis acid $SbF_5$ (one of the strongest known) displaces the Lewis acid $[AsF_2]^+$ from the Lewis base $F^-$.

(e) **EtOH readily dissolves in pyridine.** A Lewis acid/base complex formation reaction between EtOH (the acid) and py (the base) produces the adduct EtOH–py, which is held together by the kind of dative bond that we refer to as a hydrogen bond.

**6.3    Select the compound with the named characteristic. (a) Strongest Lewis acid:    BF$_3$, BCl$_3$, or BBr$_3$?** The simple argument that more electronegative substituents lead to a stronger Lewis acid does not work in this case. Boron tribromide is observed to be the strongest Lewis acid of these three compounds. The shorter boron-halogen bond distances in $BF_3$ and $BCl_3$ than in $BBr_3$ are believed to lead to stronger halogen-to-boron $p$-$p$ $\pi$ bonding (see **Section 6.6**). According to this explanation, the acceptor orbital (empty $p$ orbital) on boron is involved to a greater extent in $\pi$ bonding in $BF_3$ and $BCl_3$ than in $BBr_3$, the Lewis acidity of $BF_3$ and $BCl_3$ are diminished relative to $BBr_3$.

**BeCl$_2$ or BCl$_3$?** Boron trichloride is expected to be the stronger Lewis acid of the two for two reasons. The first reason, which is more obvious, is that the oxidation number of boron in $BCl_3$ is +3 while it is only +2 for beryllium in $BeCl_2$. The second reason has to do with structure. The boron atom in $BCl_3$ is only three-coordinate, leaving a vacant site to which a Lewis base can coordinate. Since $BeCl_2$ is polymeric, each beryllium atom is four-coordinate, and some Be–Cl bonds must be broken before adduct formation can take place.

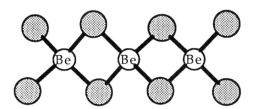

A piece of the infinite linear chain structure of $BeCl_2$. Each Be atom is four-coordinate, and each Cl atom is two-coordinate. The polymeric chains are formed by extending this piece to the right and to the left

**B($n$-Bu)$_3$ or B($t$-Bu)$_3$?**   The Lewis acid with the unbranched substituents, B($n$-Bu)$_3$, is the stronger of the two because, once the complex is formed, steric repulsions between the substituents and the Lewis base will be less than with the bulky, branched substituents in B($t$-Bu)$_3$.

**(b) More basic toward BMe$_3$: NMe$_3$ or NEt$_3$?**   These two bases have nearly equal basicities towards the proton in aqueous solution or in the gas phase.   Steric repulsions between the substituents on the bases and the proton are negligible, since the proton is very small.   However, steric repulsions between the substituents on the bases and *molecular* Lewis acids like BMe$_3$ are an important factor in complex stability, and so the smaller Lewis base NMe$_3$ is the stronger in this case.

**2-Me-py or 4-Me-py?**   As above, steric factors favor complex formation with the smaller of two bases that have nearly equal Brønsted basicities.   Therefore, 4-Me-py is the stronger base toward BMe$_3$, since the methyl substituent in this base cannot affect the strength of the B–N bond by steric repulsions with the methyl substituents on the Lewis acid.

**6.4   Which of the following reactions have K$_{eq}$ > 1? (a) R$_3$P–BBr$_3$ + R$_3$N–BF$_3$ $\rightleftharpoons$ R$_3$P–BF$_3$ + R$_3$N–BBr$_3$.** From the discussion in **Section 6.3**, we know that phosphines are softer bases than amines.   So, to determine the position of this equilibrium, we must decide which Lewis acid is softer, since the softer acid will preferentially form a complex with a soft base than with a hard base of equal strength.   Boron tribromide is a softer Lewis acid than BF$_3$, a consequence of the relative hardness and softness of the respective halogen substituents.   Therefore, the equilibrium position for this reaction will lie to the left, the side with the soft-soft and hard-hard complexes, so the equilibrium constant is less than 1.   In general, it is found that soft substituents (or ligands) lead to a softer Lewis acid than for the same central element with harder substituents.

**(b)** $SO_2 + Ph_3P-HOCMe_3 \rightleftharpoons Ph_3P-SO_2 + HOCMe_3$. In this reaction, the soft Lewis acid sulfur dioxide displaces the hard acid *t*-butyl alcohol from the soft base triphenylphosphine. The soft–soft complex is favored, so the equilibrium constant is greater than 1.

The adduct formed between triphenylphosphine and sulfur dioxide.

**(c)** $CH_3HgI + HCl \rightleftharpoons CH_3HgCl + HI$. Iodide is a softer base than chloride, an example of the general trend that elements later in a group are softer than their progenors. The soft acid $CH_3Hg^+$ will form a stronger complex with iodide than with chloride, while the hard acid $H^+$ will prefer chloride, the harder base. Thus, the equilibrium constant is less than 1.

**(d)** $[AgCl_2]^- (aq) + 2CN^- (aq) \rightleftharpoons [Ag(CN)_2]^- (aq) + 2Cl^- (aq)$. Cyanide is a softer and generally stronger base than chloride (**Table 6.1**). Therefore, cyanide will displace the relatively harder base from the soft Lewis acid $Ag^+$. The equilibrium constant is greater than 1.

**6.5**    **Choose between the two basic sites in Me₂NPF₂.** The phosphorus atom in $Me_2NPF_2$ is the softer of the two basic sites, so it will bond more strongly with the softer Lewis acid $BH_3$. See the answer to **End-of-Chapter Exercise 4(a)**, above, to review the reason why $BH_3$ is expected to be a softer Lewis acid than $BF_3$. Recall that $BH_3$ exists as dimeric molecules (see **Section 3.2**).

**6.6**    **Why does Me₃N form a relatively weak complex with BMe₃?** Since trimethylamine is the strongest Brønsted base in the gas phase, the reason that it does not form the most stable complex with trimethylboron can only be steric repulsions between the methyl substituents on the acid and those on the base.

**6.7    Discuss the relative basicity of: (a) Acetone and DMSO.** Since both $E_B$ and $C_B$ are larger for DMSO than for acetone, DMSO is the stronger base regardless of how hard or how soft the Lewis acid is. The ambiguity for DMSO is that both the oxygen atom and sulfur atom are potential basic sites.

**(b) Me₂S and DMSO.** Dimethylsulfide has a $C_B$ value that is two-and-a-half times larger than that for DMSO, while its $E_B$ value is only one quarter that for DMSO. Thus, depending on the $E_A$ and $C_A$ values for the Lewis acid, either base could be stronger. For example, DMSO is the stronger base toward $BF_3$, while $SMe_2$ is the stronger base toward $I_2$. This can be predicted by calculating the $\Delta H$ of complex formation for all four combinations:

$$DMSO\text{–}BF_3, \quad \Delta H = -[(20.21)(2.76) + (3.31)(5.83)] = -75.3 \text{ kJ mol}^{-1}$$
$$SMe_2\text{–}BF_3, \quad \Delta H = -[(20.21)(0.702) + (3.31)(15.26)] = -64.8 \text{ kJ mol}^{-1}$$

$$DMSO\text{–}I_2 \quad \Delta H = -[(4.184)(2.76) + (4.184)(5.83)] = -17.6 \text{ kJ mol}^{-1}$$
$$SMe_2\text{–}I_2 \quad \Delta H = -[(4.184)(0.702) + (4.184)(15.26)] = -32.6 \text{ kJ mol}^{-1}$$

The stable complex of DMSO and boron trifluoride

The stable complex of dimethyl sulfide and iodine

**6.8    Write balanced equations for the preparation of: (a) KBPh₄ using KPh.** Displace the hard acid $K^+$ from the soft base $Ph^-$ ($C_6H_5^-$) with the softer acid triphenylboron:

$$KPh + BPh_3 \rightarrow KBPh_4$$

Note that this is a displacement reaction, even though it seems as though $K^+$ is still associated on the right. The basic site on $Ph^-$ is the carbon atom without a hydrogen substituent. In KPh, the potassium ion is strongly associated with this carbon atom, while in the salt $KBPh_4$, it is the boron atom that is bonded to this carbon atom.

**(b) KPF₆ using PF₅.** Displace $F^-$ from $K^+$ with the stronger acid $PF_5$:

$$KF + PF_5 \rightarrow KPF_6$$

**(c) $Me_4Sn$ using MeI.** Perform a double displacement (metathesis) reaction between a Grignard reagent (an adduct of hard $Mg^{2+}$ and soft $CH_3^-$) and tetrachlorostanane (an adduct of Sn(IV), softer than $Mg^{2+}$, and $Cl^-$, harder than the methyl carbanion):

$$MeI + Mg \rightarrow MeMgI$$

$$4MeMgI + SnCl_4 \rightarrow 4MgCl_2 + Me_4Sn$$

**(d) $(SiH_3)_2O$ from $SiH_3I$.** Another metathesis reaction is called for here. Our starting material is $SiH_3I$, a complex of hard Si(IV) and the soft iodide ion. What is needed is a metal oxide metathesis reagent capable of exchanging oxide for iodide, namely the oxide of a soft metal Lewis acid such as $Ag^+$:

$$2SiH_3I + Ag_2O \rightarrow 2AgI + (SiH_3)_2O$$

**(e) $KSO_2F$ from $SO_2$.** Displace $F^-$ from $K^+$ with the stronger acid $SO_2$:

$$KF + SO_2 \rightarrow KSO_2F$$

**(f) $BF_3$ using $AsF_3$.** Perform a metathesis reaction, transfering $F^-$ from soft As(III) to hard B(III) while transfering $Cl^-$, a softer base than $F^-$, from boron to arsenic:

$$AsF_3 + BCl_3 \rightarrow AsCl_3 + BF_3$$

Note that $BBr_3$ or $BI_3$ could be used as well as $BCl_3$.


**6.9    Write a balanced equation for the dissolution of $SiO_2$ by HF.** The metathesis of two hard acids with two hard bases occurs as an equilibrium is established between solid, insoluble $SiO_2$ and soluble $H_2SiF_6$:

$$SiO_2 + 6\,HF \rightleftharpoons 2\,H_2O + H_2SiF_6 \quad \text{or}$$

$$SiO_2 + 4HF \rightleftharpoons 2H_2O + SiF_4$$

**6.10   Write a balanced equation to explain the foul odor of damp $Al_2S_3$.**
The foul odor suggests a volatile compound is formed when $Al_2S_3$ comes in contact with water.   The only volatile species that could be present, other than odorless water, is $H_2S$, which has the characteristic odor of rotten eggs. Thus, an equilibrium is established between two hard acids, Al(III) and $H^+$, and the bases $O^{2-}$ and $S^{2-}$:

$$Al_2S_3 + 3\,H_2O \rightleftharpoons Al_2O_3 + 3\,H_2S$$

**6.11   Describe the solvent properties that would: (a) Favor displacement of $Cl^-$ by $I^-$ from an acid center.**   Since in this case we have no control over the hardness or softness of the acid center, we must do something else that will favor the acid–iodide complex over the acid–chloride complex.   If we choose a solvent that decreases the activity of chloride relative to iodide, we can shift the following equilibrium to the right:

$$\text{acid–}Cl^- + I^- \rightleftharpoons \text{acid–}I^- + Cl^-$$

Such a solvent should interact more strongly with chloride (i.e. form an adduct with chloride) than with iodide.   Thus, the ideal solvent properties in this case would be *weak*, *hard*, and *acidic*.   It is important that the solvent be a weak acid, since otherwise the activity of both halides would be rendered negligible.   An example of a suitable solvent is anhydrous HF.   Another suitable solvent is $H_2O$

**(b)  Favor basicity of $R_3As$ over $R_3N$.**   In this case we wish to enhance the basicity of the soft base trialkylarsine relative to the hard base trialkylamine.   We can decrease the activity of the amine if the solvent is a hard acid, since the solvent–amine complex would then be less prone to dissociate than the solvent–arsine complex.   Alcohols such as methanol or ethanol would be suitable.

The Lewis acid/base complex of a trialkylamine (a hard base) and an alcohol (a hard acid).

(c) **Favor acidity of $Ag^+$ over $Al^{3+}$.** If we review the answers to parts (a) and (b) of this Exercise, a pattern will emerge. In both cases, a hard acid solvent was required to favor the reactivity of a soft base. In this part of the Exercise, we want to favor the acidity of a soft acid, so logically a solvent that is a hard base is suitable. Such a solvent will "tie up" (i.e. decrease the activity of) the hard acid $Al^{3+}$ relative to the soft acid $Ag^+$. An example of a suitable solvent is diethyl ether. Another suitable solvent is $H_2O$.

(d) **Promote the reaction $2FeCl_3 + ZnCl_2 \rightleftharpoons Zn^{2+} + 2[FeCl_4]^-$.** Since $Zn^{2+}$ is a softer acid than $Fe^{3+}$, a solvent that promotes this reaction will be a softer base than $Cl^-$. The solvent will then displace $Cl^-$ from the Lewis acid $Zn^{2+}$, forming $[Zn(solv)_x]^{2+}$. The solvent must also have an appreciable dielectric constant, since ionic species are formed in this reaction. A suitable solvent is acetonitrile, MeCN.

**6.12 Why are acidic solvents useful for the preparation of reactive cations and basic solvents useful for the preparation of reactive anions?** The cationic species referred to, $I_2^+$ and $Se_8^{2+}$, are intrinsically acidic, since they bear a positive charge and are not coordinatively saturated. All but the weakest bases could potentially react with them to form complexes. Since it is the free cationic species that are frequently the targets of study, and not complexes of them, the base strength of any base present must be minimized. Therefore, to insure that the reaction mixture contains only the weakest of bases, we must use the most strongly acidic solvent possible. In the case of $SbF_5/HSO_3F$, the most basic species in solution would be $HSO_3F$! By the same token, anionic species such as those referred to, $S_4^{2-}$ and $Pb_9^{2-}$, can only be studied in strongly basic solvents, because even the weakest of acids might form complexes with them.

**6.13 Describe the solvent properties that would allow: (a) Amphoteric behavior of $AsF_5$.** To exhibit amphoterism, $AsF_5$ must be capable of forming complexes with Lewis acids *and* Lewis bases. This will only be realized if it is dissolved in a solvent that does not form a complex with it. Such a solvent must therefore be neutral. A good example of a relatively neutral yet somewhat polar solvent is dichloromethane. The reason that it is referred to as *relatively* neutral is that it has been found that the chlorine atoms of dichloromethane are very weakly basic.

**(b)   High solubility of BF₃.**   A basic solvent would be ideal for generating concentrated solutions of boron trifluoride, since the enthalpy of solution would contain, among other terms, the favorable enthalpy of solvent–BF₃ adduct formation.   Benzene is a good solvent for BF₃ for just this reason:   it uses part of its $\pi$ electron cloud as a base to form a Lewis acid/base adduct with BF₃.   Another good solvent for BF₃ is diethyl ether.

**6.14   Propose a mechanism for the acylation of benzene.**   The mechanism described in **6.6** involves the abstraction of $Cl^-$ from $CH_3COCl$ by $AlCl_3$ to form $AlCl_4^-$ and the Lewis acid $CH_3CO^+$ (an acylium cation).   This cation then attacks benzene to form the acylated aromatic product.   An alumina surface, such as the partially dehydroxylated one or the fully dehydroxylated one shown in **Figures 6.7(b)** and **(c)**, respectively, would also provide Lewis acidic sites that could abstract $Cl^-$, as shown in the figure below:

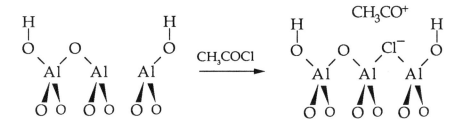

**6.15   Why does Hg(II) occur only as HgS?**   Mercury(II) is a soft Lewis acid, and so is found in nature only combined with soft Lewis bases, the most common of which is $S^{2-}$.   Sulfide can readily and permanently abstract $Hg^{2+}$ from its complexes with harder bases in ore forming geological reaction mixtures.   Zinc(II), which exhibits borderline behavior, is harder and forms stable compounds (i.e. complexes) with hard bases such as $O^{2-}$, $CO_3^{2-}$, and silicates as well as with $S^{2-}$.   The particular ore that is formed with $Zn^{2+}$ depends on factors such as the relative concentrations of the competing bases, etc.

**6.16    What is the connection between the high boiling point of water and its effectiveness for dissolving salts?**  The higher boiling point of water is due to the network of strong hydrogen bonds that exists in liquid $H_2O$.  Since each water molecule can be the donor for two hydrogen bonds and the acceptor for two, the network is stronger for water than for ammonia, which cannot accept as many hydrogen bonds per molecule.  Furthermore, because of the greater electronegativity difference between O and H than between N and H, the hydrogen bonds in $H_2O$ are stronger than the hydrogen bonds in $NH_3$.  The trends in solubilities are indeed related to this.  When a solute disrupts the structure of liquid water, stronger intermolecular interactions are broken than when the same solute disrupts the structure of liquid ammonia.  Only when the forces between water molecules and solute are strong enough to compensate for this loss, will appreciable solubility be observed.  The strong ion-dipole forces generated when salts dissolve in water are adequate compensation, but the weaker dipole-dipole or dipole-induced dipole forces generated when neutral organic compounds dissolve in water are not.

**6.17    Are the *f*-block elements hard?**  Compare these elements with Hg(II) and Zn(II), discussed in **End-of-Chapter Exercise 15**.  Since the trivalent lanthanides and actinides are found as complexes with hard oxygen bases (i.e. silicates) and not with soft bases such as sulfide, they must be hard.  Since they are found *exclusively* as silicates, they must be considered very hard, unlike the borderline behavior of Zn(II).

**6.18    Which is the stronger acid, $SiO_2$ or $CO_2$?**  $CaCO_3(s) + SiO_2(s) \rightarrow [CaSiO_3]_n + CO_2(g)$.  Silicon dioxide is a stronger Lewis acid than $CO_2$, since it abstracts an oxide ion, $O^{2-}$, from $CO_2$ in this important geochemical reaction.  The net reaction, which is a displacement reaction, is the following:

$$CO_3^{2-} + SiO_2 \rightarrow [SiO_3^{2-}]_n + CO_2$$

**6.19    Identify the acids and bases in: $TiO_2 + Na_2S_2O_7 \rightarrow Na_2SO_4 + TiO(SO_4)$.**  The net reaction is the displacement of the Lewis base $SO_4^{2-}$ from $S_2O_7^{2-}$, its adduct with the Lewis acid $SO_3$, by the strong Lewis base $O^{2-}$, as shown below:

$$O^{2-} \ + \ \left[ \substack{O \\ O{\diagdown}S{\diagup}O{\diagdown}S{\diagup}O \\ {\diagup}{\diagdown} \ \ {\diagup}{\diagdown} \\ O \ \ O \ \ O \ \ O} \right]^{2-} \longrightarrow 2 \ \left[ \substack{O \\ O{\diagdown}S{\diagup}O \\ {\diagup}{\diagdown} \\ O \ \ O} \right]^{2-}$$

**6.20   AsF₅ and [AsF₆]⁻.**   An antimony atom has five valence electrons. There are five As–F bonds in AsF₅ compound. Therefore, the central arsenic atom has 10 electrons in its valence shell in the compound (five single bonds), and the structure is a trigonal bipyramid. There are six As–F bonds in [AsF₆]⁻. In this ion, the central arsenic atom has 12 electrons in its valence shell, and the structure is an octahedron.

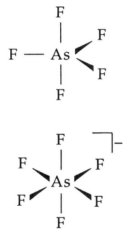

A trigonal bipyramid has a $C_3$ axis as its principal axis, three $C_2$ axes that are perpendicular to it, and two sets of mirror planes, a single horizontal plane of symmetry ($\sigma_h$) and three vertical planes of symmetry ($\sigma_v$). These symmetry elements define the $D_{3h}$ point group. An octahedral ion such as [AsF₆]⁻, with its three $C_4$ and four $C_3$ axes, has $O_h$ symmetry.

**X₃B–NH₃.**   The point group of this Lewis acid/base adduct depends on the conformation of the molecule. In either the eclipsed or staggered conformations, there are three vertical planes of symmetry besides the $C_3$ axis, and so the point group for these two conformers is $C_{3v}$. However, if the dihedral angle between the B–X and N–H bonds is anything other than 60° or 0°, the three mirror planes are lost and the point group is $C_3$. The projections below represent views down the B–N bond for the three types of conformers:

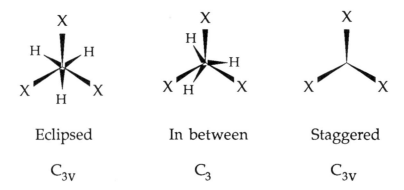

| Eclipsed | In between | Staggered |
| $C_{3v}$ | $C_3$ | $C_{3v}$ |

**Al$_2$Cl$_6$ (g).** This molecule has three mutually perpendicular $C_2$ axes, a center of symmetry, and three mutually perpendicular planes of symmetry. These symmetry elements define the $D_{2h}$ point group.

# *d*-METAL COMPLEXES

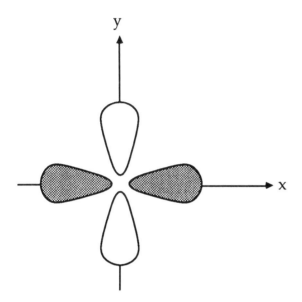

Many properties of complexes of the *d*-block metals can be under-stood on the basis of the relative energies of the five *d* orbitals that are part of the valence shell of a particular metal atom or ion. These include their structure, their spectra, their magnetic behavior, and some aspects of their thermodynamic and kinetic reactivity. Ligands in an octahedral complex are frequently placed on the *x, y*, and *z* axes. This results in the metal $d_{x^2-y^2}$ orbital, shown at the left, becoming a metal-ligand σ antibonding MO.

## SOLUTIONS TO IN-CHAPTER EXERCISES

**7.1    Identifying isomers.**  The two square-planar isomers of [PtBrCl(PR$_3$)$_2$] are shown below.  The NMR data indicate that isomer A is the *trans* isomer since the two trialkylphosphine ligands occupy opposite corners of the square plane.  Isomer B is the *cis* isomer.  Note that the two phosphine ligands in the *trans* isomer are related by symmetry elements that this $C_{2v}$ molecule possesses, namely the $C_2$ axis (the Cl–Pt–Br axis) and the $\sigma_v$ mirror plane that is perpendicular to the molecular plane. Therefore, they exhibit the same chemical shift in the $^{31}P$ NMR spectrum of this compound. The two phosphine ligands in the *cis* isomer are *not* related by the $\sigma$ mirror plane that this $C_s$ molecule possesses. Since they are chemically nonequivalent, they give rise to separate groups of $^{31}P$ resonances.

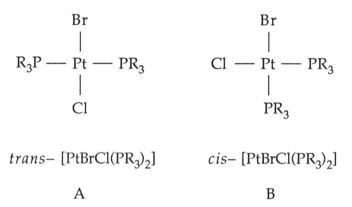

$$trans-\ [PtBrCl(PR_3)_2]$$

$$cis-\ [PtBrCl(PR_3)_2]$$

A                                    B

**7.2    Sketches of the *mer* and *fac* isomers of [Co(gly)₃].** The anion of glycine is an unsymmetrical bidentate ligand  (it has a neutral amine nitrogen donor atom    and    a    negatively charged  carboxylate  oxygen donor  atom).    In  the *fac* isomer,  the  three  possible N–Co–N bond angles are ~90°, while in the *mer* isomer, two N–Co–N bond angles are ~90° and one is 180°.  If we imagine that the complex is a sphere, the three N atoms in the *mer* isomer  lie  on  a *meridian* of the sphere (the largest circle that can be drawn on the surface of the sphere).  In contrast, the three N atoms in the *fac* isomer form one of the eight triangular *faces* of the [Co(gly)₃] octahedron.

*fac*– [Co(gly)₃]                    *mer*– [Co(gly)₃]

**7.3    Give formulas corresponding to the following names: (a) *Cis*–diaqua-dichloroplatinum(II).** As with most Pt(II) complexes, this complex is square-planar. The *cis* prefix indicates adjacent positions for the two aqua and two chloro ligands. The formula is *cis*-[PtCl$_2$(OH$_2$)$_2$]. Note the order used in naming complexes of the *d*-block metals (ligands are listed in alphabetical order, then the metal ion is listed).

**(b) Diammine*tetrakis*(isothiocyanato)chromate(III).** This is an octahedral complex of Cr(III) The *ate* suffix indicates an overall negative charge. Two NH$_3$ molecules and four NCS$^-$ anions are bonded to the Cr(III) ion. The NCS$^-$ ligands use their N atoms to bond to Cr(III) (-isothiocyanato-, not -thiocyanato-). The formula is [Cr(NCS)$_4$(NH$_3$)$_2$]$^-$. Note that a complex with this composition can exist as two structural isomers, *cis*-[Cr(NCS)$_4$(NH$_3$)$_2$]$^-$ or *trans*--[Cr(NCS)$_4$(NH$_3$)$_2$]$^-$.

**(c) *Tris*(ethylenediamine)rhodium(III).** This is also an octahedral complex. The metal ion is Rh(III), there are three symmetric bidentate ligands, and the complex is not anionic (-rhodium(III), not -rhodate(III)). Since the accepted abbreviation for ethylenediamine (H$_2$NCH$_2$CH$_2$NH$_2$) is en, the formula may be written as [Rh(en)$_3$]$^{3+}$.

**7.4    Which of the following are chiral? (a) *Cis*-[CrCl$_2$(ox)$_2$]$^{3-}$.** Drawings of two mirror images of this complex are shown below. They are not superimposable, and therefore represent two enantiomers. Therefore, this complex is chiral. Note that it does not possess an $S_n$ axis, only a single $C_2$ axis. The point group of both enantiomers is $C_2$, so they are dissymmetric, not asymmetric.

**(b) *Trans*-[CrCl$_2$(ox)$_2$]$^{3-}$.** Drawings of two mirror images of this complex are also shown below. They *are* superimposable, and therefore do not represent two enantiomers but only a single isomer. Since this complex is achiral, it must possess at least one $S_n$ axis. In fact, it possesses three different σ planes of symmetry, each of which is an $S_1$ axis.

*cis*                                                    *trans*

(c) **[RhH(CO)(PR₃)₂].** This is a complex of Rh(I), which is a $d^8$ metal ion. As discussed in **Section 7.1**, four-coordinate $d^8$ complexes of Period 5 and Period 6 metal ions are almost always square planar, and [RhH(CO)(PR₃)₂] is no exception. The bulky PR₃ ligands are *trans* to one another. This compound has $C_{2v}$ symmetry, with the $C_2$ axis coincident with the four collinear atoms H, Rh, C, and O. The two $\sigma_v$ symmetry planes prevent this complex from being chiral. *Any* planar complex cannot be chiral, whether it is square planar, trigonal planar, etc.

**7.5    What is the $d$ electron configuration of [Mn(NCS)₆]⁴⁻?** Since each iso-thiocyanate ligand has a single negative charge, the oxidation state of the manganese ion is II. Since Mn(II) is $d^5$, there are two possibilities for an octahedral complex, low-spin ($t_{2g}^5$), with one unpaired electron, or high-spin ($t_{2g}^3 e_g^2$), with five unpaired electrons. The observed magnetic moment of 6.06 $\mu_B$ is close to the spin-only value for five unpaired electrons, $(5 \times 7)^{1/2} =$ 5.92 $\mu_B$. Therefore, this complex is high-spin and has a $t_{2g}^3 e_g^2$ configuration.

**7.6    Account for the variation in $\Delta H_L$ of octahedral difluorides.** If it were not for ligand field stabilization energy, LFSE, MF₂ lattice enthalpies would increase from Mn(II) to Zn(II). This is because the decreasing ionic radius, which is due to the increasing $Z_{eff}$ as we cross through the $d$ block from left to right, leads to decreasing M–F separations. Therefore, we expect that $\Delta H_L$ for MnF₂ (2780 kJ mol⁻¹) will be smaller than $\Delta H_L$ for ZnF₂ (2985 kJ mol⁻¹).

In addition, as discussed for aqua ions and oxides, we expect additional LFSE for these compounds. From **Table 7.3**, we see that LFSE = 0 for Mn(II), $0.4\Delta_o$ for Fe(II), $0.8\Delta_o$ for Co(II), $1.2\Delta_o$ for Ni(II), and 0 for Zn(II). The deviations of the observed values from the straight line connecting Mn(II) and Zn(II) are not quite in the ratio $0.4 : 0.8 : 1.2$, but note that the deviation for Fe(II) is smaller than that for Ni(II).

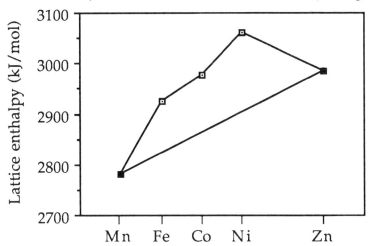

**7.7    Account for an increase of the second formation constant with respect to the first.** Normally, when a square-planar complex binds two ligands (one above and one below the square plane) $K_1 > K_2$. In this case, since the order is reversed, a structural and/or electronic rearrangement must be occuring during the second step. Since the metal is coordinated to the fairly rigid porphyrin macrocycle, a major structural change is unlikely. However, an electronic rearragement at the $d^6$ Fe(II) metal center does occur, as evidenced by the change from five-coordinate *high-spin* ($t_{2g}^4 e_g^2$) to six-coordinate *low-spin* ($t_{2g}^6$). The change in ligand field stabilization energy (LFSE) on going from high-spin $d^6$ to low-spin $d^6$ is $0.4\Delta_o - 2.4\Delta_o = -2.0\Delta_o$. This negative enthalpy change makes $\Delta G$ for the second ligand binding equilibrium more negative than it would otherwise be, and $K_2 > K_1$.

## SOLUTIONS TO END-OF-CHAPTER EXERCISES

**7.1    Trends of the *d*-block elements.** The figure below shows the Period 4 *d*-block metals as +2 cations, along with their $d^n$ configuration and group number.  Below each element is a point group classification indicating whether the $M^{2+}$ cation forms tetrahedral $(T_d)$ or square planar $(D_{4h})$ $MX_4^{2-}$ tetrahalo complexes.  There is no point group classification for $Sc^{2+}$, because this is a very uncommon oxidation state for scandium ($Sc^{2+}$ exists in the gas phase, but in condensed phases the oxidation state of scandium is almost always $Sc^{3+}$).  For Period 4 $d^8$ and $d^9$ metal ions, the square planar conformation is not as common as it for Period 5 and Period 6 analogues.  For $Ni^{2+}$ both $T_d$ and $D_{4h}$ tetrahalo complexes have been isolated, depending on the halide ion and the counterion used.  For four-coordinate $Cu^{2+}$ complexes, the $D_{4h}$ conformation is common.  The last row of information in the figure shows relative formation constants for the binding of fluoride ion, a hard ligand, to the $M^{2+}$ cations.  The series shown, an increase from $Mn^{2+}$ to $Cu^{2+}$ and then a decrease to $Zn^{2+}$, is known as the Irving-Williams series and is observed for a wide range of hard ligands.

| 3 | 4 | 5 | 6 | 7 | 8 | 9 | 10 | 11 | 12 |
|---|---|---|---|---|---|---|---|---|---|
| $Sc^{2+}$ $d^1$ | $Ti^{2+}$ $d^2$ | $V^{2+}$ $d^3$ | $Cr^{2+}$ $d^4$ | $Mn^{2+}$ $d^5$ | $Fe^{2+}$ $d^6$ | $Co^{2+}$ $d^7$ | $Ni^{2+}$ $d^8$ | $Cu^{2+}$ $d^9$ | $Zn^{2+}$ $d^{10}$ |
| — | $T_d$ | $T_d$ | $T_d$ | $T_d$ | $T_d$ | $T_d$ | both | $D_{4h}$ | $T_d$ |
| $K_{F^-} < K_{F^-} < K_{F^-} < K_{F^-} < K_{F^-} > K_{F^-}$ |||||||||| |

**7.2   Draw structures of complexes that contain the ligands (a) en, (b) ox, (c) tren, and (d) edta.**  Generic drawings of octahedral complexes of metal M for **(a)**, **(b)**, and **(c)** are shown below.  In **(a)**, the bidentate ligand ethylenediamine (en = $H_2NCH_2CH_2NH_2$) takes up two coordination sites. For clarity, the carbon atoms of the ethylene bridge are not explicitly shown. The five membered ring that is formed is *not* planar — one carbon atom is above and one is below the plane formed by M and the two N atoms (see **Structural Drawing 17**).  In **(b)**, the bidentate ligand oxalate dianion (ox = $C_2O_4^{2-}$) also takes up two coordination sites.  Once again, the carbon atoms of the ligand are shown simply as vertices.  Due to the delocalized $\pi$ system of the ligand, the five membered ring in this case *is* planar.  In **(c)**, the tetradentate ligand triethylenetetramine (tren = $N(CH_2CH_2NH_2)_3$) takes up four coordination sites.  Both the carbon and hydrogen atoms on the ethylene bridges have been omitted for clarity.  For a drawing of a complex of ethylenediamine-tetraacetate (edta = $(O_2CCH_2)_2NCH_2CH_2N(CH_2CO_2)_2{}^{4-}$) see **Structural Drawing 18**.

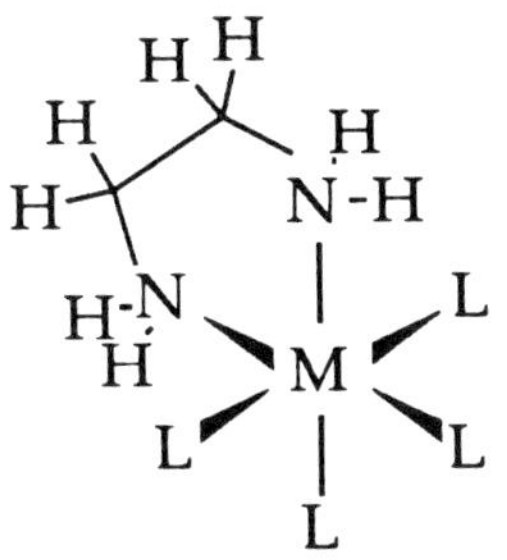

**(a)** $ML_4(en)$ 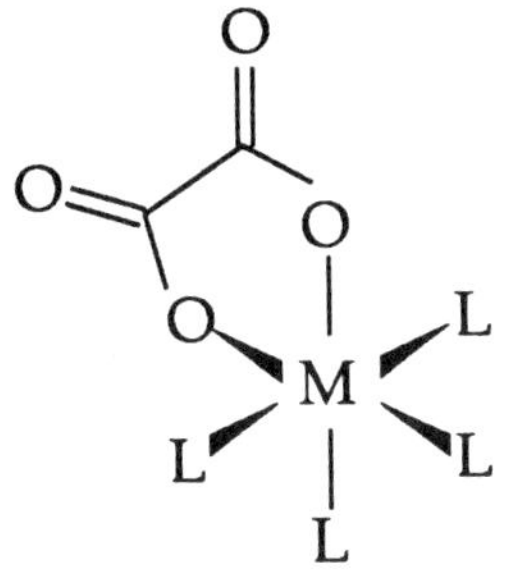      **(b)** $ML_4(ox)$ 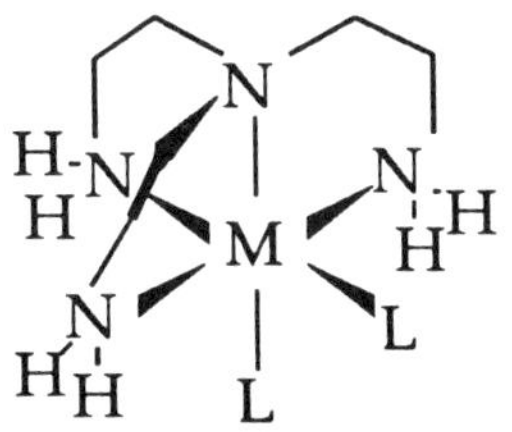      **(c)** $ML_2(tren)$

**7.3**    **Name and draw structures.**  **(a)** *cis*-$[CrCl_2(NH_3)_4]^+$**.** This complex contains (i) two $Cl^-$ ligands = dichloro, (ii) four $NH_3$ ligands = tetraammine (not tetraamine or tetramine or tetrammine), and (iii) $Cr^{3+}$ in a complex that is not an anion = chromium(III). Putting (i), (ii), and (iii) together, the name of this complex is *cis*-tetraamminedichlorochromium(III). Note that the ligands are named in alphabetical order: tetra*a*mmine before di*c*hloro. The structure is shown below.

    **(b)** *Trans*-$[Cr(NCS)_4(NH_3)_2]^-$**.** This complex contains (i) four N-bonded $NCS^-$ ligands = tetraisothiocyanato, (ii) two $NH_3$ ligands = diammine, and (iii) $Cr^{3+}$ in a complex that is an anion = chromate(III). Therefore, the name of this complex is *trans*-diammine-tetraisothiocyanatochromate(III). Note the alphabetical ordering of the ligands. The structure is shown below.

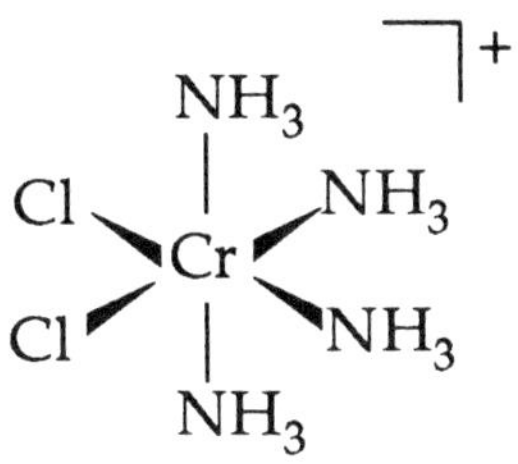

*cis*-$[CrCl_2(NH_3)_4]^+$ 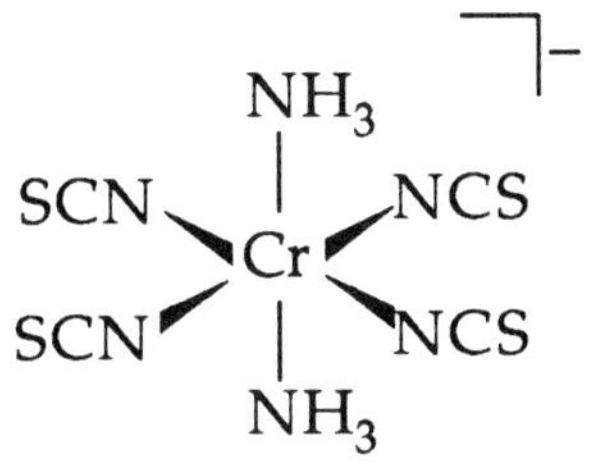      *trans*-$[Cr(NCS)_4(NH_3)_2]^-$ 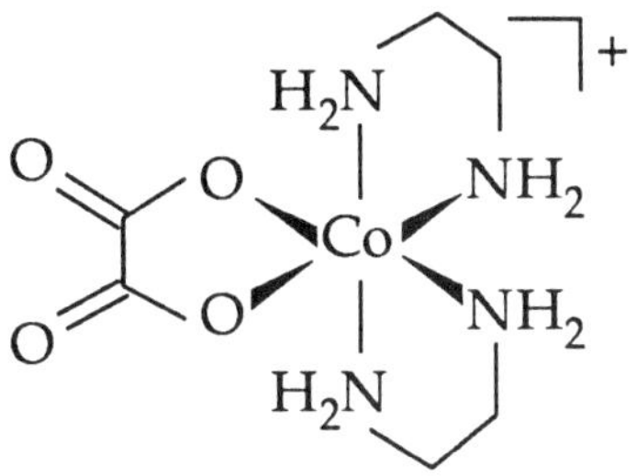      $[Co(ox)(en)_2]^+$

**(c)** **[Co(ox)(en)₂]⁺.** This complex contains (i) one oxalate dianion ligand $(C_2O_4{}^{4-})$ = oxalato, (ii) two ethylenediamine ligands = *bis*(ethylenediamine), and (iii) $Co^{3+}$ in a complex that is not an anion = cobalt(III).    Therefore,  the   name   of   this  complex   is *bis*(ethylenediamine)oxalatocobalt(III).  The structure is shown above.  Note that we do not need to use *cis-* or *trans-* for a complex of this composition, since neither of the two types of bidentate ligands in the complex can span two *trans* positions.

**7.4**   **Draw structures of:  (a)  Pentaamminechlorocobalt(III) chloride; (b) Hexaaquairon(III) nitrate.**  The structures of these two compounds are shown below.

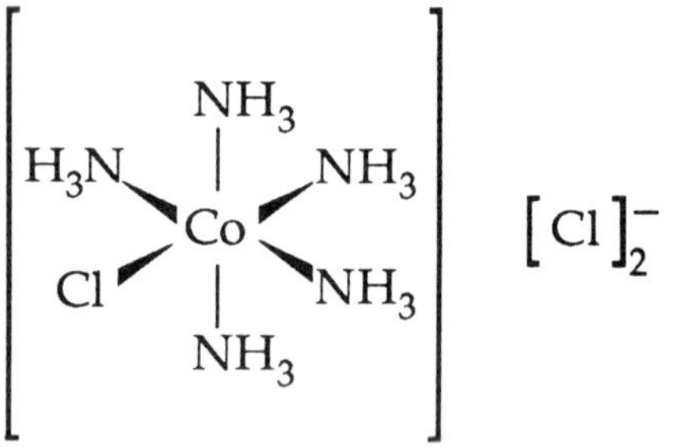

[CoCl(NH₃)₅][Cl]₂ 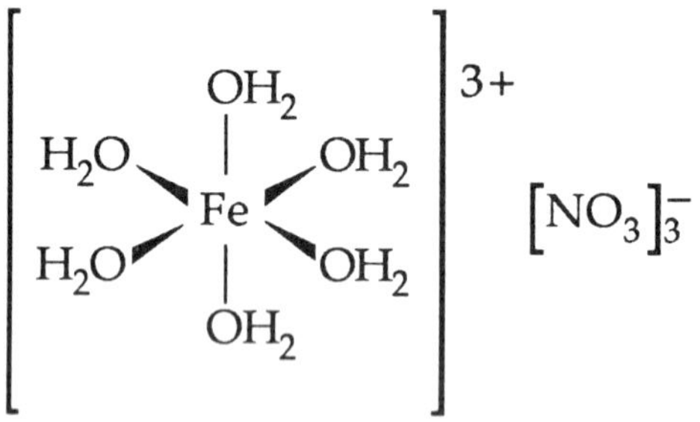                    [Fe(OH₂)₆][NO₃]₃

**(c)** ***Cis*-dichloro*bis*(ethylenediamine)ruthenium(II) ion;**    **(d)** **µ-Hydroxo*bis*(pentaamminechromium(III)) chloride.**  These structures are shown below.

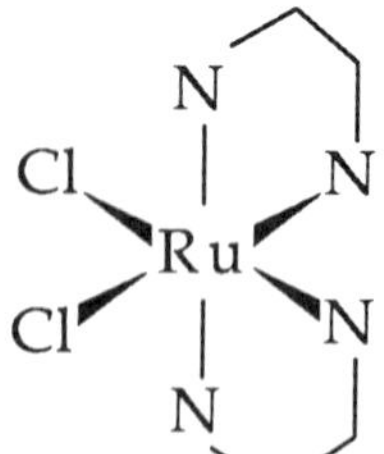

*cis*-[RuCl₂(en)₂]                    [(H₃N)₅Cr(µ-OH)Cr(NH₃)₅][Cl]₅  (L = NH₃)

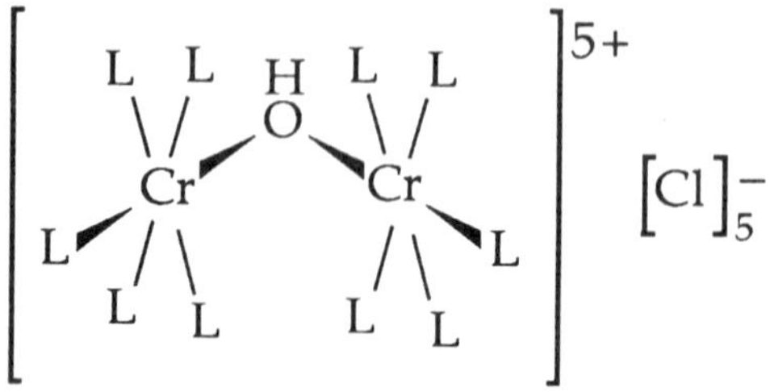

**7.5** **Draw all possible isomers of: (a) Octahedral [RuCl₂(NH₃)₄]; (b) Square-planar [IrH(CO)(PR₃)₂].** For each compound, *cis* and *trans* isomers are possible, as shown below.

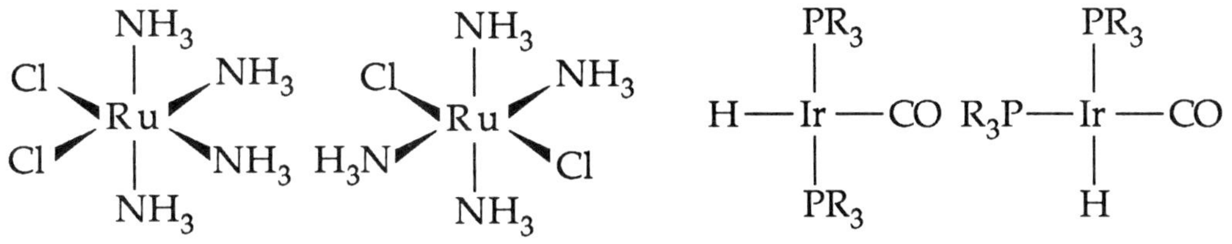

$$\text{Octahedral } [RuCl_2(NH_3)_4]; \qquad \text{Square-planar } [IrH(CO)(PR_3)_2]$$

cis       trans       trans       cis

**(c) Tetrahedral [CoCl₃(OH₂)]; (d) Octahedral [CoCl₂(NH₃)₂(en)]⁺.** There is only one isomer for **(c)**, but one *cis* and two *trans* isomers for **(d)**, as shown below (the + charge for the **(d)** isomers is not shown).

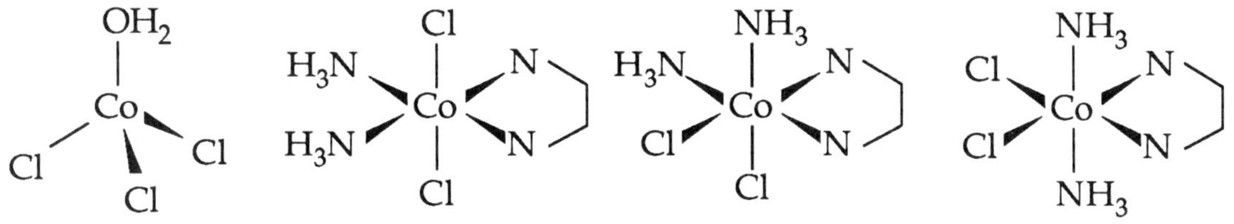

*trans*(1)       *cis*       *trans*(2)

**7.6** **Which of the following complexes are chiral? (a) [Cr(ox)₃]³⁻.** All octahedral *tris*(bidentate ligand) complexes are chiral, since they can exist as either a right hand or a left hand propeller, as shown in the drawings of the two nonsuperimposable mirror images (the oxalate ligands are shown in abbreviated form and the −3 charge on the complexes has been omitted).

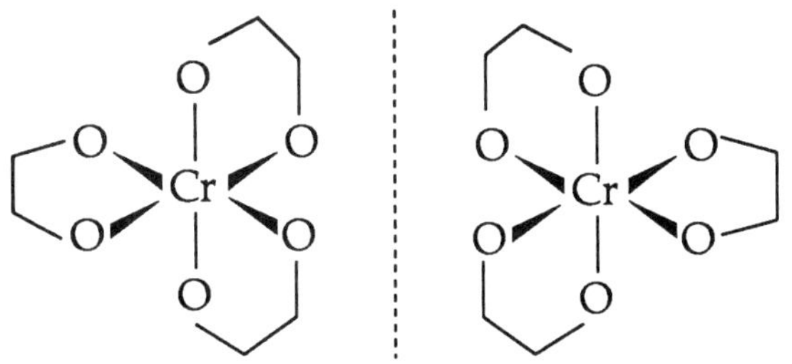

**(b) *Cis*-[PtCl₂(en)].** This is a four-coordinate complex of a Period 6 $d^8$ metal ion, so it is undoubtedly square-planar. As discussed above in the answer to **In-Chapter Exercise 4(c)**, any planar complex contains at least one plane of symmetry and must be achiral. In this case the five-membered chelate ring formed by the ethylenediamine ligand is not planar, so, strictly speaking, the complex is not planar. It *can* exist as two enantiomers, depending on the conformation of the chelate ring, as shown below. However, the conformational interconversion of the ethylene linkage is extremely rapid, so the two enantiomers cannot be separated.

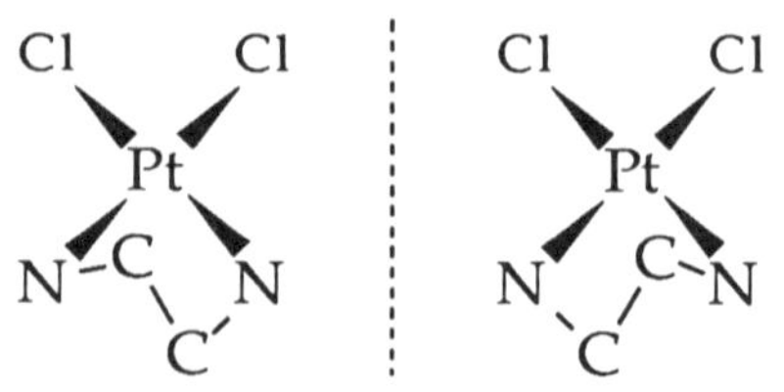

*d* and *l* [PtCl₂(en)] 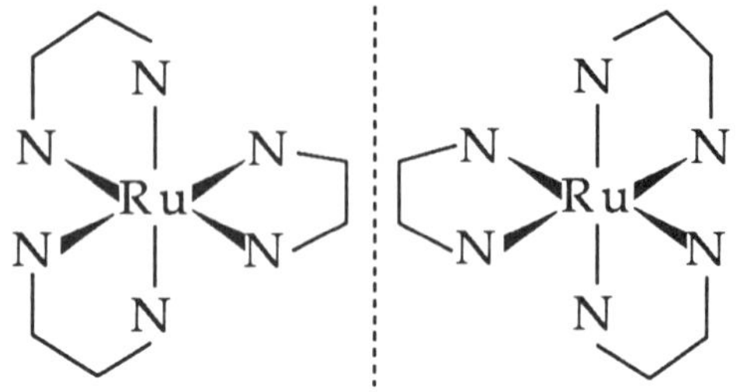                          Δ and Λ [Ru(bipy)₃]³⁺

   (c) *Cis*-[RhCl₂(NH₃)₄]⁺. This complex has $C_{2v}$ symmetry so it is not chiral. The $C_2$ axis is coincident with the bisector of the Cl–Rh–Cl bond angle, one $\sigma_v$ plane is coincident with the plane formed by the Rh atom and the two Cl atoms, and the other $\sigma_v$ plane is perpendicular to the first.

   (d) [Ru(bipy)₃]²⁺. As stated in the answer to part (a), above, all octahedral *tris*(bidentate ligand) complexes are chiral, and this one is no exception. The two nonsuperimposable mirror images are shown above (the bipyridine ligands are shown in abbreviated form and the +3 charge on the complex has been omitted).

   (e) [Co(edta)]⁻. A drawing of one of the enantiomers of this complex is shown in **Structure 18**. If we interchange the top and bottom oxygen atoms in this drawing, along with their respective acetate arms, we will form the other enantiomer.

   (f) *Fac*-[Co(NO₂)₃(dien)]. If we ignore the conformations of the five-membered chelate rings, this complex, shown below, is not chiral. The plane perpendicular to the page that includes the central diethylenetriamine N atom and the Co atom is a symmetry plane for this complex (since the complex possesses $C_s$ symmetry, this plane is the only symmetry element present).

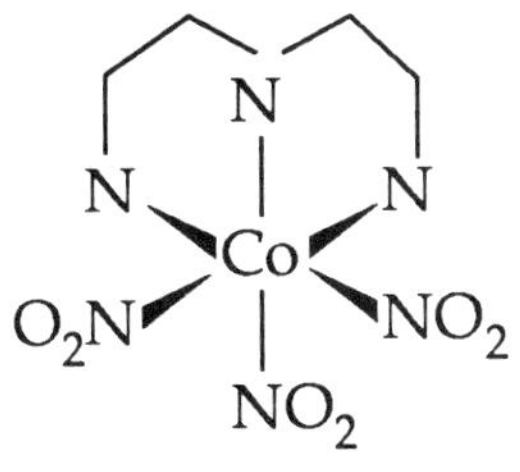

*fac* 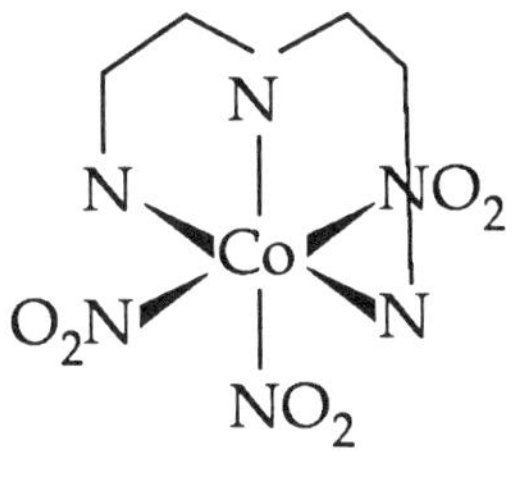                                       *mer*

If we take into account the various conformations of the ethylene linkages, the stereo-isomer possibilities are much more complicated. As explained in the text for en, the ethylene linkages undergo *rapid* conformational interconversion

**(g) *Mer*-[Co(NO₂)₃(dien)].** As above, if we ignore the conformations of the five-membered chelate rings, this complex, which is shown above, is not chiral. The plane that contains the Co atom and all three dien N atoms is a symmetry plane of the complex, as is the plane that contains the Co atom and all three nitrite N atoms. Ignoring the chelate rings, this complex belongs to the $C_{2v}$ point group.

**7.7    Deduce the structures of pink CoCl₃·5NH₃·H₂O and purple CoCl₃·5NH₃.** Since the pink complex rapidly gives 3 mol AgCl on titration with AgNO₃ solution, the Cl⁻ ions cannot be coordinated to the Co³⁺ ion, which forms inert complexes. Instead, they must be counterions, and the five ammonia molecules and the water molecule must comprise the six ligands of this octahedral complex. The purple solid, on the other hand, does not contain any water, so one of the Cl⁻ ions must now be coordinated to Co³⁺. The structures of the two complex ions are shown below. The name of the pink compound is pentaammineaquacobalt(III) chloride (the two neutral ligands are listed in alphabetical order, ammine before aqua). The name of the purple compound is pentaamminechlorocobalt(III) chloride. Dots are used in formulas for complexes where the structure is unknown or unspecified.

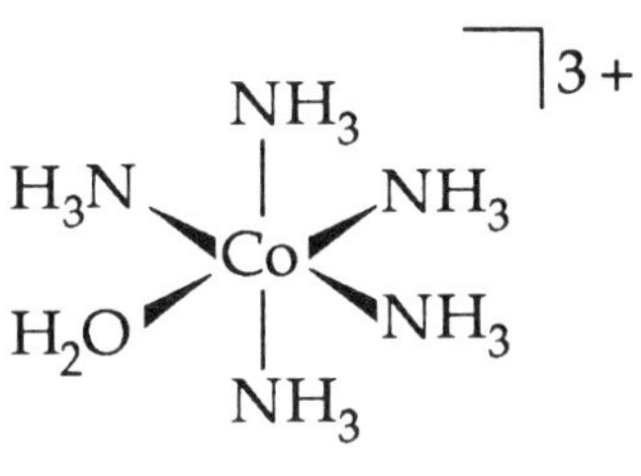

pink [Co(NH₃)₅(OH₂)]³⁺ 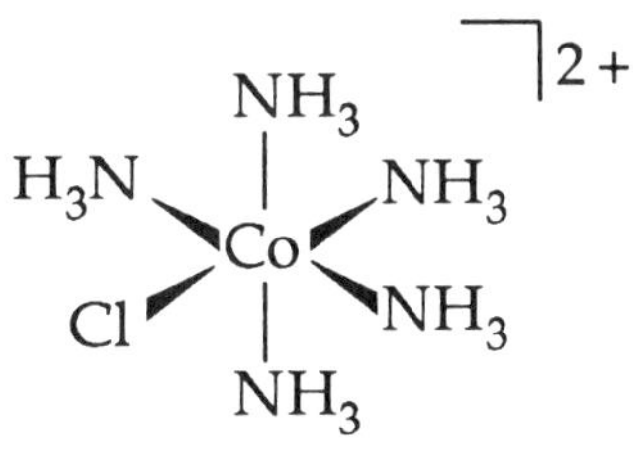              purple [CoCl(NH₃)₅]²⁺

**7.8    Deduce the structures of blue $CrCl_3 \cdot 6H_2O$ and green $CrCl_3 \cdot 5H_2O$.** Since the conductivity data suggests that the three $Cl^-$ ions are counterions (i.e. they are not coordinated to the $Cr^{3+}$ ion), the blue compound must contain the hexaaquachromium(III) ion, shown below. The green compound has a lower conductivity, suggesting that one or more of the $Cl^-$ ions are coordinated to $Cr^{3+}$, leaving fewer $Cl^-$ ions to be counterions. Since the formula contains five water molecules, it is reasonable to presume that the green compound contains the pentaaquachlorochromium(III) ion, shown below.

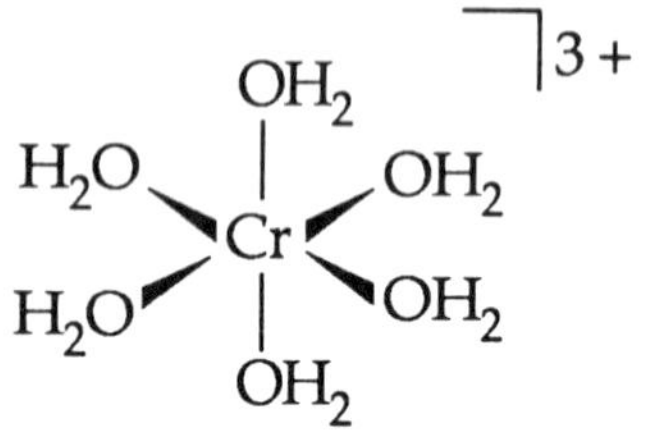

blue complex 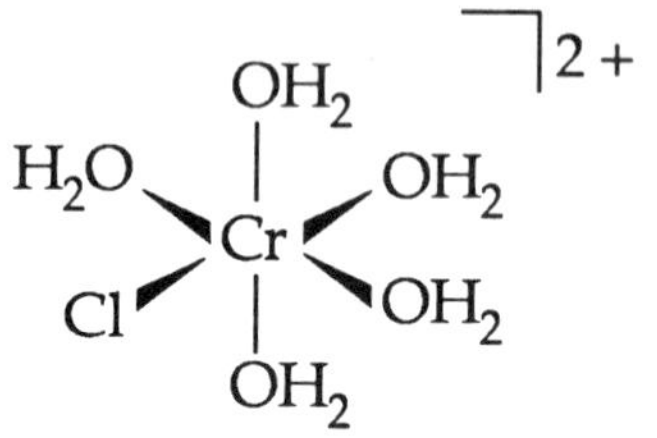                                        green complex

**7.9    Name and draw the structure of the diaqua $[Pt(NH_3)_2(OH_2)_2]^{2+}$ complex.** The observation that this complex does not react with ethylenediamine, which is unable to span two opposite positions in a square planar complex, suggests that this is the *trans* isomer. The name of the complex ion is diamminediaquaplatinum(II) (alphabetical order requires *ammine* before *aqua*). Its structure is shown below and to the left.

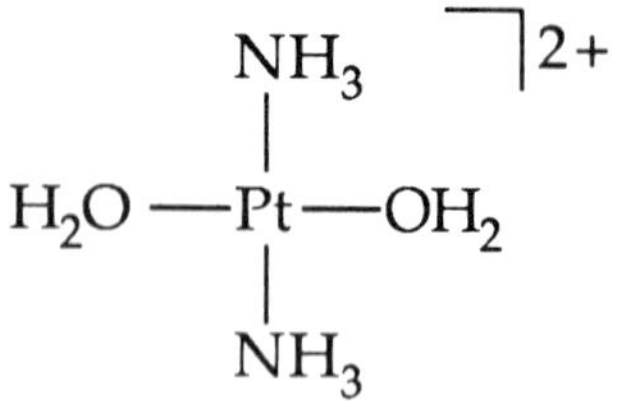

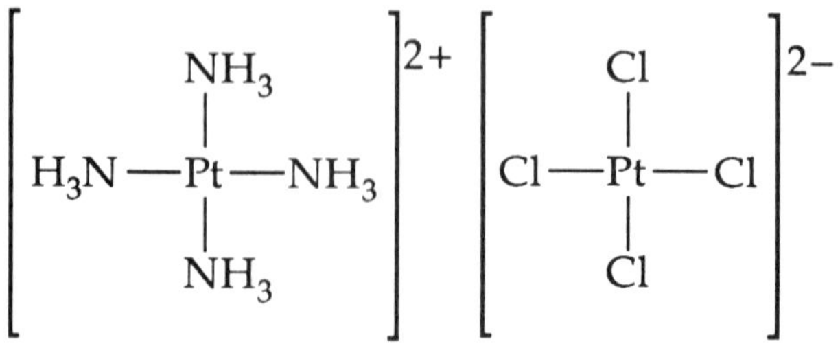

**7.10    Give the structure and name of the parent $PtCl_2 \cdot 2NH_3$ compound.** The metathesis reaction with $AgNO_3$ suggests the presence of $[Pt(NH_3)_4]^{2+}$ and $[PtCl_4]^{2-}$:

$$[Pt(NH_3)_4][PtCl_4] + AgNO_3 \rightarrow [Pt(NH_3)_4][NO_3] \text{ and } Ag[PtCl_4]$$

This formulation would also explain the lack of formation of any AgCl. The name of this compound is tetraammineplatinum(II) tetrachloroplatinate(II), and the structures of the two square-planar complex ions that comprise this compound are shown above and to the right.

**7.11    Give the structures of Jensen's alkylphosphine and -arsine complexes.** Since these compounds do not have a dipole moment, they must be *trans* isomers. As discussed in **Section 2.4**, $C_{2v}$ molecules like the *cis* isomers of these compounds will have dipole moments whereas $D_{2h}$ molecules, like the *trans* isomers, will not. The structures of *trans*-[PtCl$_2$(PR$_3$)$_2$] and *trans*-[PtCl$_2$(AsR$_3$)$_2$] are shown at the right.

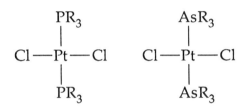

**7.12    Which of the following complexes obey the 18-electron rule?    (a) [Cu(NH$_3$)$_4$]$^{2+}$.** The copper ion in this complex, which is coordinated to only neutral $NH_3$ ligands, is $Cu^{2+}$, a $d^9$ metal ion. Since each of the four ammonia molecules is a two-electron donor, the total number of electrons in the valence shell of the copper ion is 9 + 4 x 2 = 17. One of the hallmarks of 18-electron complexes is that they are diamagnetic (completely filled bonding and nonbonding sets of orbitals). Mononuclear $Cu^{2+}$ complexes are always paramagnetic, with one unpaired electron regardless of geometry, and can never obey the 18-electron rule.

**(b) [Fe(CN)$_6$]$^{4-}$.** The iron ion in this complex, which is coordinated to six negatively charged $CN^-$ ion ligands, is $Fe^{2+}$, a $d^6$ metal ion. Since each of the six cyanide ions is a two-electron donor, the total number of electrons in the valence shell of the copper ion is 6 + 6 x 2 = 18. This complex does obey the 18-electron rule.

**(c) [Fe(CN)$_6$]$^{3-}$.** This complex has a –3 charge instead of the –4 charge of the $Fe^{2+}$ complex in **(b)**, so it clearly contains $Fe^{3+}$. This being so, the complex has one fewer electron, namely 17, that the complex in **(b)**. The complex $[Fe(CN)_6]^{3-}$ does not obey the 18-electron rule.

**(d)** $[Cr(NH_3)_6]^{3+}$. The chromium ion in this complex, which is coordinated to six neutral $NH_3$ ligands, is $Cr^{3+}$, a $d^3$ metal ion. Since each of the six ammonia molecules is a two–electron donor, the total number of electrons in the valence shell of the copper ion is $3 + 6 \times 2 = 15$. This complex does not obey the 18-electron rule.

**(e)** $[Cr(CO)_6]$. Since the CO ligands are neutral and the complex is not charged, the central metal in this complex is $Cr^0$, which is a $d^6$ metal atom. (There is not any confusion about the $4s$ vs. $3d$ energies for metal atoms when they are surrounded by ligands — in complexes of metal atoms all of the metal's valence electrons are in its $d$ subshell.) Since each of the CO ligands is a two–electron donor, the total number of electrons in the valence shell of the chromium atom is $6 + 6 \times 2 = 18$. Many metal carbonyl complexes, including this one and the one in **(f)**, below, obey the 18-electron rule.

**(f)** $[Fe(CO)_5]$. The iron atom in this complex contributes eight electrons to the total number of electrons, since $Fe^0$ is a $d^8$ metal atom. Therefore, the total number of electrons is $8 + 5 \times 2 = 18$. See **(e)**, above, for more details.

**7.13** **Draw bonding and antibonding $\sigma$ orbitals for *trans*-$[W(CO)_4(PR_3)_2]$.** This complex has $D_{4h}$ symmetry. The two linear combinations of P atom $\sigma$ orbitals are shown below. Spherical $s$-type orbitals are used in the two linear combinations, which have $A_{1g}$ and $A_{2u}$ symmetries, but the P atom $\sigma$ orbitals could also be hybrid orbitals. The central W atom $d_{z^2}$ orbital also has $A_{1g}$ symmetry, and so can form bonding and antibonding combinations with the $A_{1g}$ combination of phosphorus orbitals.

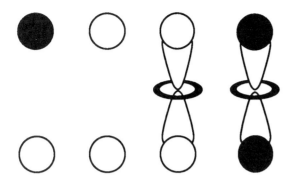

The combination on the far left is the $A_{2u}$ linear combination of phosphorus $\sigma$ orbitals. Moving to the right, we have the $A_{1g}$ combination of phosphorus *s* orbitals, followed by the bonding and antibonding molecular orbitals formed by taking + and – combinations of the $A_{1g}$ phosphorus combination and the metal $d_{z^2}$ orbital

**7.14  $d^n$, number of unpaired electrons, and LFSE.** **(a) $[Co(NH_3)_6]^{3+}$.** Since the $NH_3$ ligands are neutral, the cobalt ion in this octahedral complex is $Co^{3+}$, which is a $d^6$ metal ion. Ammonia is in the middle of the spectrochemical series, but since the cobalt ion has a +3 charge, this is a strong field complex and hence is low-spin, with $S = 0$ and no unpaired electrons (the configuration is $t_{2g}^6$). The LFSE is $6(0.4\Delta_0) = 2.4\Delta_0$. Note that this is the largest possible value of LFSE for an octahedral complex.

**(b) $[Fe(OH_2)_6]^{2+}$.** The iron ion in this octahedral complex, which contains only neutral water molecules as ligands, is $Fe^{2+}$, which is a $d^6$ metal ion. Since water is lower in the spectrochemical series than $NH_3$ (i.e. it is a weaker field ligand than $NH_3$) *and* since the charge on the metal ion is only +2, this is a weak field complex and hence is high-spin, with $S = 2$ and four unpaired electrons (the configuration is $t_{2g}^4 e_g^2$). The LFSE is $4(0.4\Delta_0) - 2(0.6\Delta_0) = 0.4\Delta_0$. Compare this small value to the large value for the low–spin $d^6$ complex in **(a)** above.

**(c) $[Fe(CN)_6]^{4-}$.** The iron ion in this octahedral complex, which contains six negatively charged $CN^-$ ion ligands, is $Fe^{2+}$, which is a $d^6$ metal ion. Cyanide ion is a very strong field ligand, so this is a strong field complex and hence is low-spin, with $S = 0$ and no unpaired electrons. Like $[Co(NH_3)_6]^{3+}$, the configuration is $t_{2g}^6$ and the LFSE is $2.4\Delta_0$.

**(d) $[Cr(NH_3)_6]^{3+}$.** The complex contains six neutral $NH_3$ ligands, so chromium is $Cr^{3+}$, a $d^3$ metal ion. The configuration is $t_{2g}^3$, and so there are three unpaired electrons. Note that for octahedral complexes, only $d^4$-$d^7$ metal ions have the possibility of being either high-spin or low-spin. For $[Cr(NH_3)_6]^{3+}$, the LFSE = $3(0.4\Delta_o)$ = $1.2\Delta_o$. (For $d^1$-$d^3$, $d^8$, and $d^9$ metal ions in octahedral complexes, only one spin state is possible).

(e) **$[W(CO)_6]$.** Like the chromium atom in $Cr(CO)_6$ (see the answer to **End-of-Chapter Exercise 7.12(e)**, above), the W atom in this octahdral complex is $d^6$. Since CO is such a strong field ligand (it is even higher in the spectrochemical series than $CN^-$), $W(CO)_6$ is a strong field complex and hence is low-spin, with no unpaired electrons (the configuration is $t_{2g}^6$). The LFSE = $6(0.4\Delta_o)$ = $2.4\Delta_o$.

(f) **Tetrahedral $[FeCl_4]^-$.** The iron ion in this complex, which contains four negatively charged $Cl^-$ ion ligands, is $Fe^{3+}$, which is a $d^5$ metal ion. All tetrahedral complexes are high–spin, since $\Delta_T$ is much smaller than $\Delta_o$ ($\Delta_T = (4/9)\Delta_o$, if the metal ion, the ligands, and the metal-ligand distances are kept constant) so for this complex $S$ = 5/2 and there are five unpaired electrons. The configuration is $e^2t^3$. The LFSE is $2(0.6\Delta_T) - 3(0.4\Delta_T)$ = 0.

(g) **Tetrahedral $[Ni(CO)_4]$.** The neutral CO ligands require that the metal center in this complex is $Ni^0$, which is a $d^{10}$ metal atom. Regardless of geometry, complexes of $d^{10}$ metal atoms or ions will never have any unpaired electrons and will always have LFSE = 0, and this complex is no exception.

**7.15 What factors account for the ligand field strength of different ligands?** It is clear that $\pi$-acidity cannot be a requirement for a position high in the spectrochemical series, since $H^-$ is a very strong field ligand but is not a $\pi$-acid (it has no *low energy* acceptor orbitals of local $\pi$ symmetry). However, ligands that are very strong $\sigma$-bases will increase the energy of the $e_g$ orbitals in an octahedral complex relative to the $t_{2g}$ orbitals. Thus, there are two ways for a complex to develop a large value of $\Delta_o$, by possessing ligands that are $\pi$–acids *or* by possessing ligands that are strong $\sigma$-bases (of course some ligands, like $CN^-$, exhibit both $\pi$-acidity and moderately strong $\sigma$-basicity). A class of ligands that are also very high in the spectrochemical series are alkyl anions, $R^-$ (e.g. $CH_3^-$). These are not $\pi$-acids but, like $H^-$, are very strong bases (see **Table 5.3**).

**7.16  Comment on the lattice enthalpies for CaO, TiO, VO, MnO, FeO, CoO, and NiO.** As in the answer to **In-Chapter Exercise 7.6**, there are two factors that lead to the values given in this question and plotted below: decreasing ionic radius from left to right across the *d*-block, leading to a general increase in $\Delta H_L$ from CaO to NiO, and LFSE, which varies in a more complicated way for high-spin metal ions in an octahedral environment, increasing from $d^0$ to $d^3$, then decreasing from $d^3$ to $d^5$, then increasing again from $d^5$ to $d^8$, then decreasing again from $d^8$ to $d^{10}$. The straight line through the black squares is the trend expected for the first factor, the decrease in ionic radius (the last black square is not a data point, but simply the extrapolation of the line between $\Delta H_L$ values for CaO and MnO, both of which have LFSE = 0). The deviations of $\Delta H_L$ values for TiO, VO, FeO, CoO, and NiO from the straight line is a manifestation of the second factor, the nonzero values of LFSE for $Ti^{2+}$, $V^{2+}$, $Fe^{2+}$, $Co^{2+}$, and $Ni^{2+}$. We shall find in **Chapter 18** that TiO and VO have considerable metal-metal bonding, and this factor also contributes to their stability.

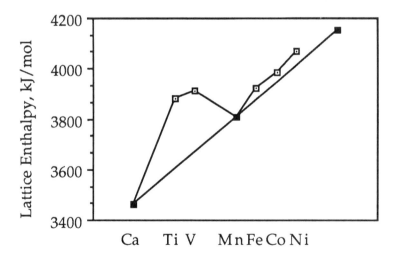

**7.17  Predict the structure of $[Cr(OH_2)_6]^{2+}$.** The main consequence of the Jahn-Teller theorem (See **Section 7.6**) is that a nonlinear molecule or ion with an orbitally degenerate ground state is not as stable as a distorted version of the molecule or ion if the distortion removes the degeneracy. The high-spin $d^4$ complex $[Cr(OH_2)_6]^{2+}$ has the configuration $t_{2g}^3 e_g^1$, which is orbitally  degenerate since the single $e_g$ electron can be in either the $d_{z2}$ or the

$d_{x2-y2}$ orbital. Therefore, by the Jahn-Teller theorem, the complex should not have $O_h$ symmetry. A tetragonal distortion, whereby two *trans* metal-ligand bonds are elongated and the other four are shortened, removes the degeneracy. This is the most common distortion observed for octahedral complexes of high-spin $d^4$, low-spin $d^7$, and $d^9$ metal ions, all of which possess $e_g$ degeneracies and exhibit measureable Jahn-Teller distortions. The predicted structure of the $[Cr(OH_2)_6]^{2+}$ ion, with the elongation of the two *trans* Cr–O bonds shown greatly exaggerated, is shown at the right.

**7.18    Explain the asymmetry in the $e_g \leftarrow t_{2g}$ visible transition for Ti$^{3+}$ (*aq*).** It is suggested that we explain this observation using the Jahn-Teller theorem. The ground state of Ti(OH$_2$)$_6^{3+}$, which is a $d^1$ complex, is not one of the configurations that usually leads to an observable Jahn-Teller distortion (the three main cases are listed in the answer to **End-of-Chapter Exercise 7.17**, above). However, the electronic excited state of Ti(OH$_2$)$_6^{3+}$ has the configuration $t_{2g}^0e_g^1$, and so the excited state of this complex possesses an $e_g$ degeneracy. Therefore, the "single" electronic transition is really the superposition of two transitions, one from an $O_h$ ground state ion to an $O_h$ excited state ion, and a lower energy transition from an $O_h$ ground state ion to a lower energy distorted excited state ion (probably $D_{4h}$). Since this two transitions have slightly different energies, the unresolved superimposed bands result in an asymmetric absorption peak.

**7.19    Explain which ligand forms the more stable complex with Ni$^{2+}$ (*aq*).** **(a) CH$_3$OH or CH$_3$NH$_2$.** These are both neutral monodentate ligands. Since methylamine is a considerably stronger base than methanol, Ni(CH$_3$NH$_2$)$^{2+}$ (*aq*) is a stronger complex than Ni(CH$_3$OH)$^{2+}$ (*aq*).

**(b) (CH$_3$)$_2$CHCH$_2$NH$_2$ or NH$_2$CH(CH$_3$)CH$_2$NH$_2$.** The nitrogen atoms in these two ligands have about the same basicity. Furthermore, the two ligands are about the same size, so steric effects are nearly the same.

However, the first ligand is monodentate while the second is bidentate. Due to the chelate effect, the second ligand will form a stronger complex with $Ni^{2+}$ than will the first.

(c) **$NH_3$ or $NF_3$?** Ammonia is a much stronger Brønsted base and Lewis base than $NF_3$. It is therefore a considerably stronger ligand than $NF_3$. As a consequence, $NH_3$ will form a stronger complex with $Ni^{2+}$ than will $NF_3$.

(d) **Three $NH_2CH_2CH_2NH_2$ or two $NH_2CH_2CH_2NHCH_2CH_2NH_2$?** The nitrogen atoms in these two ligands have about the same basicity. Furthermore, steric effects are nearly the same. However, the first ligand is bidentate while the second is tridentate. Due to the fact that more chelate rings are formed with the second ligand than with the first, the second ligand will form the stronger complex with $Ni^{2+}$.

**7.20 Account for the generalizations that $F^-$ and $O^{2-}$ are used to stabilize metal centers with high oxidation numbers while phosphines and CO are used to stabilize those with low oxidation numbers.** There are two differences between these two types of ligands. First of all, fluoride and oxide ions are hard bases while trialkyl- and triarylphosphines and carbon monoxide are soft bases. Hard bases form more stable complexes with hard acids, and soft bases prefer soft acids. For a given metal, a high oxidation number (generally > +3) will give a hard metal center, while a low oxidation number (generally < +2) will give a soft metal center. Furthermore, fluoride and oxide ions are π-donors while phosphines and CO are π-acceptors. High oxidation states are stabilized by π-donors because this class of ligands can place needed electron density on the electron-poor metal center. Low oxidation states are stabilized by π-acceptors because this class of ligands can remove excess electron density from the electron-rich metal center. Below are shown an example of a Cr(VI) complex with hard, π-donor ligands, and a Cr(0) complex, with soft, π-acceptor ligands.

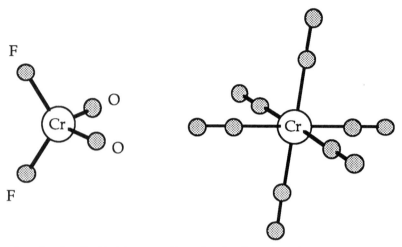

Tetrahedral $CrO_2F_2$, which contains Cr(VI).

Octahedral $Cr(CO)_6$, which contains Cr(0). The carbon atoms of the CO ligands are bound to the Cr atom.

**7.21   What is the difference between complexometric and chelatometric titrations?**   As the name implies, complexometric titrations involve the formation of a complex.   The complex could be a simple one between a solvated metal ion and a monodentate ligand, or it could involve several ligands or a polydentate ligand.   An example of a chelate complex is shown below.   A chelatometric titration results in the formation of a chelate complex, so the ligand being used is necessarily a polydentate one.   As discussed in **Section 7.7**, a single polydentate ligand with $n$ donor atoms forms a more stable complex than $n$ monodentate ligands.   In analytical titrations, it is important to achieve complex formation that is as close to 100% completion as possible, so that the amount of ligand added to reach the endpoint is a simple fraction (e.g. 1/1, 1/2, 1/3, etc.) of the amount of metal ion present.   Note that chelatometric titrations are a subset of all complexometric titrations.

This is the structure of the [Mg(edta)(H_2O)]^{2-} dianion. Hydrogen atoms have been omitted for clarity and the carbon atoms of the edta^{4-} ligand are left unlabelled. The hexadentate edta^{4-} ligand nearly surrounds the metal ion.

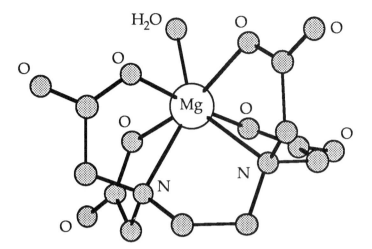

**7.22 Why is it possible to separate isomers of $[Ir(NH_3)_4Cl_2]^+$ but not of $[Cu(NH_3)_4Cl_2]$?** As discussed in **Section 7.8**, strong field octahedral $d^3$ and $d^6$ complexes are generally substitution inert, while all others are generally substitution labile. Only the former type of complexes can be separated into pure isomers. The two $d^9$ $Cu^{2+}$ isomers are labile, and rapid ligand substitution provides a mechanism for rapid isomerization. The $d^6$ $Ir^{3+}$ complexes, like octahedral $Co^{3+}$ complexes, are substitution inert. Since rapid isomerization does not occur for the iridium complexes, they can be separated and isolated as discrete *cis* and *trans* isomers.

**7.23 What is the mechanism of formation of $[Co(NH_3)_5X]^{2+}$ from $[Co(NH_3)_5OH_2]^{3+}$?** The reaction in question is ($X^- = Cl^-$, $Br^-$, $N_3^-$, and $SCN^-$):

$$[Co(NH_3)_5OH_2]^{3+} + X^- \rightarrow [Co(NH_3)_5X]^{2+} + H_2O$$

Since the rate constant does not depend strongly on the nature of the incoming ligand, it is probably *not* associative. Therefore, it is likely to be dissociative.

**7.24 If a substitution process is associative, why may it be difficult to characterize an aqua ion as labile or inert?** The classification of a complex as labile (reaction time < 1 minute) implies a judgement about its reaction with various partners. Since the rate of an associative reaction depends strongly on the entering ligand, the rates of substitution of an aqua ion may span many orders of magnitude.

$$M(OH_2)_n^{m+} + {}^*OH_2 \rightarrow M(OH_2)_{n-1}({}^*OH_2)^{m+} + OH_2$$
Could be associative and slow

$$M(OH_2)_n^{m+} + X^- \rightarrow MX(OH_2)_{n-1}^{(m-1)+} + OH_2$$
Could be associative and fast

# CHAPTER 8

# OXIDATION AND REDUCTION

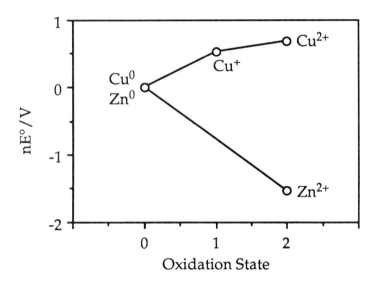

Frost diagrams, shown here for copper and zinc in aqueous acid (pH = 0), are useful representations of electrode potential data. These two graphs show that $Cu^0$ will not reduce $H^+$ ($aq$) to $H_2$, that $Zn^0$ will reduce $H^+$ ($aq$) to $H_2$ and will also reduce $Cu^{2+}$ to $Cu^0$, and that $Cu^+$ will disproportionate to $Cu^0$ and $Cu^{2+}$.

## SOLUTIONS TO IN-CHAPTER EXERCISES

**8.1** **The minimum temperature for reduction of MgO by carbon.** At about 1700 °C, the Mg, MgO line and the C, CO line cross, which means that the reactions $2Mg$ ($l$) + $O_2$ ($g$) → $2MgO$ ($s$) and $2C$ ($s$) + $O_2$ ($g$) → $2CO$ ($g$) have the same free energy change at that temperature. Thus, coupling the two reactions (i.e. subtracting the first from the second) yields the overall reaction

MgO (s) + C (s) → Mg (l) + CO (g) with $\Delta G = 0$. At 1700 °C or above, MgO can be conveniently reduced to Mg by carbon.

**8.2    The potential difference needed to reduce $TiO_2$.** The reduction of $TiO_2$ to Ti can take place electrolytically at potentials equal to or less than $E = -\Delta G/nF$, where $\Delta G$ is the free energy change for $TiO_2$ (s) → Ti (l) + $O_2$ (g). We can find approximate values for $\Delta G$ (500 °C) and $\Delta G$(1000 °C) from the Ellingham diagram shown in **Figure 8.1**, and these are −800 and −720 kJ mol$^{-1}$, respectively. Since $n = 4$ for the reduction of Ti (per mole of $O_2$),

$$E(500 \text{ °C}) = (-800 \text{ kJ mol}^{-1})(1000 \text{ J/kJ})/4(9.65 \times 10^4 \text{ C mol}^{-1}) = -2.1 \text{ J C}^{-1} = -2.1 \text{ V}$$

and

$$E(1000 \text{ °C}) = (-720 \text{ kJ mol}^{-1})/4(9.65 \times 10^4 \text{ C mol}^{-1}) = -1.9 \text{ V}$$

Note that a less negative potential is needed at higher temperature, because of the assist by the positive entropy change for the reduction.

**8.3    Can $Cr_2O_7^{2-}$ be used to oxidize $Fe^{2+}$, and would $Cl^-$ oxidation be a problem?** The standard reduction potential for the $Cr_2O_7^{2-}/Cr^{3+}$ couple is 1.38 V, so it can oxidize any couple whose reduction potential is less than 1.38 V. In general terms, if the reduction potential for Ox + $e^-$ → Red is X, then the overall potential $E$ for the (unbalanced) reaction $Cr_2O_7^{2-}$ + Red → $Cr^{3+}$ + Ox is (1.38 − X) V. Since $\Delta G = -nFE$, the reaction will have a negative free energy change as long as $E$ is positive. Thus, as long as the *reduction* potential X is less than 1.38 V, $E$ will be positive. Returning to the specific question at hand, since the reduction potential for the $Fe^{3+}/Fe^{2+}$ couple is 0.77 V, $Fe^{2+}$ will be oxidized to $Fe^{3+}$ by dichromate. Since the reduction potential for the $Cl_2/Cl^-$ couple is 1.36 V, oxidation of chloride ion is only slightly favored and in practice it is not observed to occur at an appreciable rate.

**8.4    The potential for reduction of $MnO_4^-$ to $Mn^{2+}$ at pH = 7.** The standard potential for the reduction $MnO_4^-$ (aq) + 8H$^+$ (aq) + 5$e^-$ → $Mn^{2+}$ (aq) +

$4H_2O$ (*l*) is 1.53 V (see **Table 8.1**). The pH, or $[H^+]$, dependence of the potential $E$ is given by the Nernst equation ($[MnO_4^-] = [Mn^{2+}] = 1$ M),

$$E = E° - ((0.059 \text{ V})/5)(\log(1/[H^+]^8)$$

Note that $n = 5$ for the reduction of $MnO_4^-$. Note also that the factor $0.059 \text{ V}/n$ can only be used at 25 °C. Since at pH = 7, $[H^+] = 10^{-7}$ M,

$$E = 1.53 \text{ V} - (0.0118 \text{ V})(\log(1/10^{-56}) = 1.53 \text{ V} - (0.0118 \text{ V})(56)$$

$$E = 1.53 \text{ V} - 0.66 \text{ V} = 0.87 \text{ V}$$

So, since the reduction potential is less positive, permanganate ion is a weaker oxidizing agent (i.e. it is less readily reduced) in neutral solution than at pH = 0.

**8.5    What is the potential for the rapid oxidation of Mg by $H^+$ at 25 °C and pH = 7?**  The balanced equation for the reaction in question is

$$Mg \text{ (s)} + 2H^+ \text{ (aq)} \rightarrow Mg^{2+} \text{ (aq)} + H_2 \text{ (g)}$$

The potential for this reaction can be calculated using the Nernst equation,

$$E = E° - ((0.059 \text{ V})/n)(\log Q)$$

where $n = 2$ and $Q = P(H_2)[Mg^{2+}]/[H^+]^2$. The value of $E°$ can be calculated by subtracting $E°$ for the $Mg^{2+}/Mg$ couple from $E°$ for the $H^+, H_2$ couple. The former can be found in **Appendix 4** (–2.36 V), and the latter, by definition, is 0, so $E° = 0 \text{ V} - (-2.356 \text{ V}) = 2.356 \text{ V}$. If we assume that $P(H_2) = 1$ bar and that $[Mg^{2+}] = 1$ M, then $Q = 1/[H^+]^2 = 10^{14}$ at pH = 7, and $\log Q = 14$. Therefore,

$$E = 2.356 \text{ V} - (0.0295 \text{ V})(14) = 2.356 \text{ V} - 0.413 \text{ V} = 1.943 \text{ V}$$

This value is well in excess of the 0.6 V overpotential needed in insure a rapid reaction.

**8.6    Can $Fe^{2+}$ disproportionate under standard conditions?** The disproportionation of $Fe^{2+}$ involves the reduction of one equivalent of $Fe^{2+}$ to $Fe^0$, a
net gain of two equivalents of electrons, and the concomitant oxidation of
two equivalents to $Fe^{2+}$ to $Fe^{3+}$, a net loss of two equivalents of electrons:

$$3Fe^{2+} (aq) \rightarrow Fe^0 (s) + 2Fe^{3+} (aq)$$

The value of $E$ for this reaction can be calculated by subtracting $E°$ for the
$Fe^{2+}$, Fe couple ($-0.41$ V) from $E°$ for the $Fe^{3+}/Fe^{2+}$ couple ($0.77$ V),

$$E = -0.41 \text{ V} - 0.77 \text{ V} = -1.18 \text{ V}$$

Since this potential is large and negative, the disproportionation will *not*
occur.

**8.7    The fate of $SO_2$ emitted into clouds.** We are told that the standard
reduction potential for the $SO_4^{2-}/SO_2$ couple is 0.17 V.  Since the standard
reduction potential for the $O_2/H_2O$ couple is 1.23 V, the potential for the
coupled reaction

$$2SO_2 (aq) + O_2 (g) + 2H_2O (l) \rightarrow 2SO_4^{2-} (aq) + 4H^+ (aq)$$

is $E° = 1.23$ V $- 0.16$ V $= 1.07$ V.  Since this potential is large and positive, this
reaction will be driven nearly to completion.  The aqueous solution of $SO_4^{2-}$
and $H^+$ ions precipitates as acid rain, which can have a pH as low as 2 (the pH
of rain water that is not contaminated with sulfuric or nitric acid is ~5.6).

**8.8    Calculate $E°$ for the $ClO_3^-/ClO^-$ couple.** We make use of the Latimer
diagram for chlorine in acid solution (this is given in **Appendix 4**) and the
equation:

$$E_{13} = (n_1 E_{12} + n_2 E_{23})/(n_1 + n_2)$$

In this case $E_{12}$ is the standard potential for the $ClO_3^-/ClO_2^-$ couple:

$$ClO_3^- (aq) + 2H^+ (aq) + 2e^- \rightarrow ClO_2^- (aq) + H_2O (l) \qquad E_{12} = 1.18 \text{ V}, n_1 = 2$$

and $E_{23}$ is the standard potential for the $ClO_2^-/ClO^-$ couple:

$$ClO_2^- \ (aq) \ + \ 2H^+ \ (aq) \ + \ 2e^- \rightarrow ClO^- \ (aq) \ + \ H_2O \ (l) \qquad E_{23} = 1.67 \ V, n_2 = 2$$

so $E_{13}$ = ((2)(1.18 V) + (2)(1.67 V))/4 = 1.43 V. In this case, since $n_1 = n_2$, $E_{13}$ is the simple average of $E_{12}$ and $E_{23}$, whereas in the more general case $n_1 \neq n_2$, $E_{13}$ is the *weighted* average of $E_{12}$ and $E_{23}$.

**8.9    Frost diagram for thallium in acid solution.** From the potentials given, $nE° = 0$ V for $Tl^0$ ($n = 0$), $nE° = -0.34$ V for $Tl^+$ ($n = 1$), and $nE° = 2.19$ V for $Tl^{3+}$ ($n = 3$), so a plot of $nE°/V$ vs. oxidation state $n$ is the figure at the left. Note that $Tl^+$ is stable with respect to disproportionation in aqueous acid. Note also that $Tl^{3+}$ is a strong oxidant (i.e. it is very readily reduced), since the slope of the line connecting it with either lower oxidation state is large and positive.

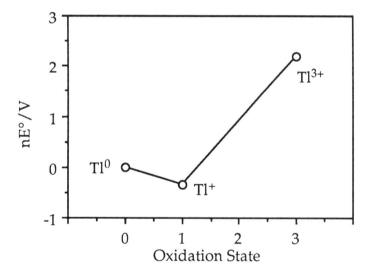

**8.10    Can $O_2$ oxidize $Mn^{2+}$ to $Mn^{3+}$ in acid solution? Can $H_2O_2$?** We can construct a Frost diagram (shown below) with just the relevant species on it, and then invoke the rule discussed in **Section 8.9**, "the oxidizing agent with the more positive slope is liable to undergo reduction." The potentials necessary for the construction of the manganese part of the diagram can be found in **Appendix 4** ($nE°$ for $Mn^{2+}$ = –2.36 V, $nE°$ for $Mn^{3+}$ = –0.86 V), while those for the oxygen diagram are given in **Example 8.9**. Since the slope of the line for the $O_2/H_2O$ couple is less positive than for the $Mn^{3+}/Mn^{2+}$ couple, oxygen cannot oxidize $Mn^{2+}$ to $Mn^{3+}$. Furthermore, the slope of the line for the $H_2O_2/H_2O$ couple is nearly the same as that for the $Mn^{3+}/Mn^{2+}$ couple, so hydrogen peroxide cannot oxidize $Mn^{2+}$ to $Mn^{3+}$ either. This is *not* an especially good application of Frost diagrams. Simple numerical calculations are easier to carry out and are more reliable.

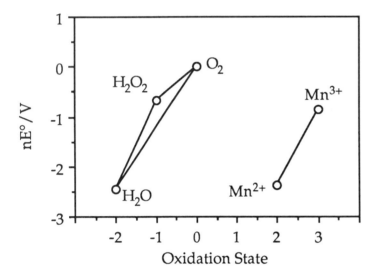

The Frost diagram for oxygen in acid solution superimposed on a portion of the Frost diagram for manganese in acid solution.

**8.11    Compare the strength of $NO_3^-$ as an oxidizing agent in acidic and basic solution.** As discussed in **Section 8.9**, the reduction of nitrate ion usually proceeds to NO, which is evolved from the solution, instead of proceeding all the way to $N_2$. If we compare the portion of the Frost diagram for nitrogen, shown below, containing the $NO_3^-/NO$ couple in acidic and in basic solution, we see that the slope for the couple in acidic solution is positive while the slope for basic solution is negative. Therefore, nitrate is a stronger oxidizing agent (i.e. it is more readily reduced) in acidic solution than in basic solution. Even if the reduction of $NO_3^-$ proceeded all the way to $N_2$, the slope of that line is still less than the slope of the line for the $NO_3^-/NO$ couple in acid solution.

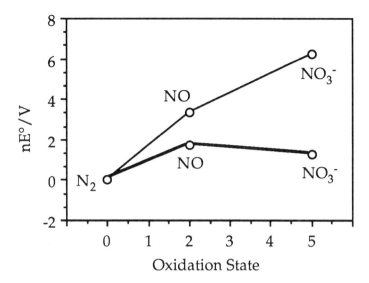

A portion of the Frost diagram for nitrogen. The points connected with the bold lines are for basic solution, and the other points are for acid solution.

**8.12   The possibility of finding $Fe_2O_3$ in a waterlogged soil.** According to **Figure 8.9**, a typical waterlogged soil (organic rich and oxygen depleted) has a pH of about 4 and a potential of about –0.1 V. If we find the point on the Pourbaix diagram for naturally occuring iron species, **Figure 8.8**, we see that $Fe_2O_3$ is not stable and $Fe^{2+}$ (*aq*) will be the predominant species. In fact, as long as the potential remained at –0.1 V, $Fe^{2+}$ remains the predominant species below pH = 8. Above pH = 8, $Fe^{2+}$ is oxidized to $Fe_2O_3$ at this potential. Note that for a potential of –0.1 V and a pH below 2, $Fe^{2+}$ is still the predominant iron species in solution, but water is reduced to $H_2$.

**8.13   Which couple has the higher $E°$, $[Ni(en)_3]^{2+}/Ni$ or $[Ni(NH_3)_6]^{2+}/Ni$? Section 8.11** has shown that complexation alters potentials. The formation of a more stable complex when the metal has the higher oxidation state, in this case $Ni^{2+}$, favors oxidation and makes the reduction potential more negative. Due to the chelate effect, ethylenediamine forms a more stable complex with $Ni^{2+}$ than does ammonia (see **Section 7.7**). Therefore, the $[Ni(NH_3)_4]^{2+}/Ni$ couple has the higher $E°$, since the $[Ni(en)_2]^{2+}/Ni$ couple favors oxidation and has the lower (or more negative) $E°$. Thermodynamically (but not mechanistically), we can think of the process occuring by decomplexation followed by reduction of the nickel(II) aqua ion to nickel metal. For either complex, the second equilibrium is exactly the same. However, the first equilibrium lies farther to the left for the ethylenediamine complex than for the ammonia complex, since the ethylenediamine complex is more stable. Therefore, the *overall* equilibrium constant is smaller for the ethylenediamine complex than for the $NH_3$ complex, resulting in a more negative potential for the ethylenediamine complex than for the $NH_3$ complex.

$$[Ni(en)_3]^{2+} (aq) \rightleftharpoons 3en + Ni^{2+} (aq) \rightleftharpoons Ni (s)$$

$$[Ni(NH_3)_6]^{2+} (aq) \rightleftharpoons 6NH_3 + Ni^{2+} (aq) \rightleftharpoons Ni (s)$$

**8.14   Identify the lanthanide ion used as follows: (a) As a strong oxidizing agent.** An ion that will serve as a strong oxidant should be readily reduced. From the data in **Table 8.2** we see that none of the +3 or +2 lanthanide ions are readily reduced — they all have large negative reduction potentials, which means that their reduction is very unfavorable relative to the reduction of $H^+$ (*aq*) to $H_2$. However, the $Ce^{4+}$ (*aq*)/$Ce^{3+}$ (*aq*) couple has a reduction potential of 1.72 V, a large and *positive* value. Therefore, $Ce^{4+}$ is a strong oxidant in aqueous solution, since it is so readily reduced.

**(b)   As a strong reducing agent.** An ion that will serve as a strong reducing agent should be readily oxidized. All of the lanthanide metals *in their elemental state* are readily oxidized (they all have large negative reduction potentials), but the only lanthanide *ion* that is readily oxidized in aqueous solution is $Eu^{2+}$ ($E° = -0.35$ V for the $Eu^{3+}$ (*aq*)/$Eu^{2+}$ (*aq*) couple). This means that $Eu^{2+}$ is a strong reducing agent in aqueous solution.

---

## SOLUTIONS TO END-OF-CHAPTER EXERCISES

**8.1   Under what conditions will Al reduce MgO?** The lines for $Al_2O_3$ and MgO on the Ellingham diagram (**Figure 8.1**) represent the change in $\Delta G°$ with temperature for the following reactions:

$$(1/3)Al (l) + O_2 (g) \rightarrow (2/3)Al_2O_3 (s) \quad \text{and} \quad 2Mg (l) + O_2 (g) \rightarrow 2MgO (s)$$

At temperatures below about 1400 °C, the free energy change for the MgO reaction is more negative than for the $Al_2O_3$ reaction. This means that under these conditions MgO is more stable with respect to its constituent elements than is $Al_2O_3$, and that Mg will react with $Al_2O_3$ to form MgO and Al. However, above about 1400 °C the situation reverses, and Al will react with MgO to reduce it to Mg with the concomitant formation of $Al_2O_3$. This is a rather high temperature, achievable in an electric arc furnace (compare the extraction of silicon from its oxide, discussed in **Section 8.1**).

**8.2    Write balanced equations, if a reaction occurs, for the following species in aerated aqueous acid.    (a)    $Cr^{2+}$.** For all of these species, we must determine whether they can be oxidized by $O_2$. The standard potential for the reduction $O_2 + 4H^+ + 4e^- \rightarrow 2H_2O$ is 1.23 V. Therefore, only redox couples with a reduction potential less positive than 1.23 V will be driven to completion to the oxidized member of the couple by the reduction of $O_2$ to $H_2O$. Since the $Cr^{3+}/Cr^{2+}$ couple has $E° = -0.424$ V, $Cr^{2+}$ *will* be oxidized to $Cr^{3+}$ by $O_2$. The balanced equation is:

$$4Cr^{2+} \ (aq) \ + \ O_2 \ (g) \ + \ 4H^+ \ (aq) \ \rightarrow \ 4Cr^{3+} \ (aq) \ + \ 2H_2O \ (l) \quad E° = 1.65 \ V$$

**(b)    $Fe^{2+}$.** Since the $Fe^{3+}/Fe^{2+}$ couple has $E° = 0.771$ V, $Fe^{2+}$ *will* be oxidized to $Fe^{3+}$ by $O_2$. The balanced equation is:

$$4Fe^{2+} \ (aq) \ + \ O_2 \ (g) \ + \ 4H^+ \ (aq) \ \rightarrow \ 4Fe^{3+} \ (aq) \ + \ 2H_2O \ (l) \quad E° = 0.46 \ V$$

**(c)    $Cl^-$.** Both of the following couples have $E°$ values, shown in parentheses, more positive than 1.23 V, so there will be no reaction with $O_2$ is mixed with aqueous chloride ion in acid solution: $ClO_4^-/Cl^-$ (1.287 V); $Cl_2/Cl^-$ (1.358 V). The appropriate equation is (NR = no reaction):

$$Cl^- \ (aq) \ + \ O_2 \ (g) \ \rightarrow \ NR$$

**(d)    HOCl.** Since the $HClO_2/HClO$ couple has $E° = 1.701$ V, HClO *will not* be oxidized to $HClO_2$ by $O_2$. The standard potential for the oxidation of HOCl by $O_2$ is $-0.47$ V.

(e) **Zn (s).** Since the $Zn^{2+}/Zn$ couple has $E° = -0.763$ V, metallic zinc *will* be oxidized to $Zn^{2+}$ by $O_2$. The balanced equation is:

$$2Zn\ (s)\ +\ O_2\ (g)\ +\ 4H^+\ (aq)\ \rightarrow\ 2Zn^{2+}\ (aq)\ +\ 2H_2O\ (l)\qquad E° = 1.99\ V$$

A competing reaction will be:

$$Zn\ (s)\ +\ 2H^+\ (aq)\ \rightarrow\ Zn^{2+}\ (aq)\ +\ H_2\ (g)\qquad E° = 0.763\ V$$

**8.3   Write the Nernst equation for:  (a)  The reduction of $O_2$.** In general terms, the Nernst equation is given by the formula:

$$E\ =\ E°\ -\ (RT/nF)\ln Q\qquad \text{where } Q \text{ is the reaction quotient}$$

For the reduction of oxygen, $O_2\ (g) + 4H^+\ (aq) + 4e^- \rightarrow 2H_2O\ (l)$,

$$Q\ =\ 1/(P(O_2)[H^+]^4)\qquad \text{and}\qquad E\ =\ E°\ +\ (RT/nF)(\ln P(O_2)\ -\ 9.2pH)$$

since $pH = -\log[H^+] = -2.3\ln[H^+]$ and $\log[H^+]^4 = 4\log[H^+]$. The potential for $O_2$ reduction at $pH = 7$ and $P(O_2) = 0.20$ bar is

$$E\ =\ 1.229\ V\ +\ (0.0148\ V)(-1.6 - 64.4)\ =\ 0.252\ V$$

   **(b)  The reduction of $Fe_2O_3$ (s).** For the reduction of solid iron(III) oxide, $Fe_2O_3\ (s) + 6H^+\ (aq) + 6e^- \rightarrow 2Fe\ (s) + 3H_2O\ (l)$,

$$Q\ =\ 1/[H^+]^6\qquad \text{and}\qquad E\ =\ E°\ -\ (RT/nF)(13.8pH)$$

since $pH = -\log[H^+] = -2.3\ln[H^+]$ and $\log[H^+]^6 = 6\log[H^+]$.

**8.4   Calculate $E$, $\Delta G°$, and $K$ for the reduction of $CrO_4^{2-}$ and $[Cu(NH_3)_4]^+$ by $H_2$ in basic solution.** We must recognize that although the two values of $E°$ are so similar, the two reactions involve different numbers of electrons, $n$, and the expressions for $\Delta G°$ and $K$ involve $n$ whereas $E$ does not:

$$\Delta G°\ =\ -nFE\qquad \text{and}\qquad \Delta G°\ =\ -RT\ln K$$

For the reduction of chromate ion:

$$\Delta G^\circ = -nFE = -(3)(9.65 \times 10^4 \text{ C mol}^{-1})(-0.11 \text{ V}) = 31.8 \text{ kJ mol}^{-1}$$

(recall that 1 V = 1 J C$^{-1}$) and since $RT = 2.48$ kJ mol$^{-1}$ at 25 °C,

$$K = \exp((-31.8 \text{ kJ mol}^{-1})/(2.48 \text{ kJ mol}^{-1})) = 2.70 \times 10^{-6}$$

For the reduction of $[Cu(NH_3)_4]^+$:

$$\Delta G^\circ = -(9.65 \times 10^4 \text{ C mol}^{-1})(-0.10) = 9.65 \text{ kJ/mol}^{-1}$$

$$\text{and } K = \exp((-9.65 \text{ kJ mol}^{-1})/(2.48 \text{ kJ mol}^{-1})) = 2.04 \times 10^{-2}$$

So, because the reductions differ by the number of electrons required (3 vs. 1), the equilibrium constants are different by a factor of about 8000.

**8.5  Using Frost diagrams. (a) What happens when $Cl_2$ is dissolved in aqueous basic solution?**  The Frost diagram for chlorine in basic solution is shown in **Figure 8.15** and is reproduced below.  If the points for $Cl^-$ and $ClO_4^-$

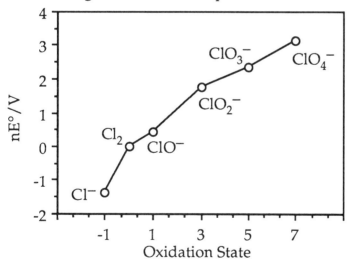

are connected by a single straight line, the point for $Cl_2$ lies above it. Therefore, $Cl_2$ is thermodynamically susceptible to disproportionation to $Cl^-$ and $ClO_4^-$ when it is dissolved in aqueous base.  In practice, the further oxidation of $ClO^-$ is slow, so a solution of $Cl^-$ and $ClO^-$ is formed when $Cl_2$ is dissolved in aqueous base.

**(b)  What happens when $Cl_2$ is dissolved in aqueous acid solution?**
The Frost diagram for chlorine in acidic solution is shown in **Figure 8.15**. If

the points for $Cl^-$ and any positive oxidation state of chlorine are connected by a single straight line, the point for $Cl_2$ lies below it (if only slightly). Therefore, $Cl_2$ will not disproportionate. However, $E°$ for the $Cl_2/Cl^-$ couple, 1.36 V, is more positive than $E°$ for the $O_2/H_2O$ couple, 1.23 V. Therefore, $Cl_2$ is thermodynamically capable of oxidizing water as follows, although the reaction is very slow:

$$Cl_2 \ (aq) \ + \ H_2O \ (l) \ \rightarrow \ 2Cl^- \ (aq) \ + \ 2H^+ \ (aq) \ + \ 1/2O_2 \qquad E° \ = \ 0.13 \ V$$

(c) **Should $HClO_3$ disproportionate in aquous acid solution?** The point for $ClO_3^-$ in acidic solution on **Figure 8.15** lies above the single straight line connecting the points for $Cl_2$ and $ClO_4^-$. Therefore, since $ClO_3^-$ is thermodynamically unstable with respect to disproportionation in acidic solution (i.e. it *should* disproportionate), the failure of it to exhibit any observable disproportionation must be due to a kinetic barrier.

**8.6   Write equations for the following reactions: (a) $N_2O$ is bubbled into aqueous NaOH solution.** Part of the Frost diagram for nitrogen in basic solution is shown below (the entire diagram is shown in **Figure 8.5**).

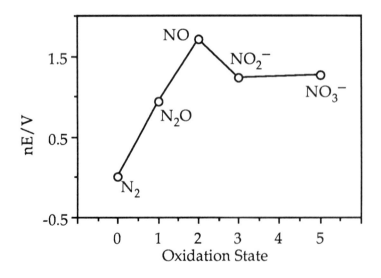

The Frost diagram for nitrogen in basic solution, showing some pertinent species.

Inspection of it shows that $N_2O$ lies above the line connecting $N_2$ and $NO_3^-$. Therefore, $N_2O$ is thermodynamically susceptible to disproportionation to $N_2$ and $NO_3^-$ in basic solution:

$$5N_2O\ (aq)\ +\ 2OH^-\ (aq)\ \rightarrow\ 2NO_3^-\ (aq)\ +\ 4N_2\ (g)\ +\ H_2O\ (l)$$

However, the redox reactions of nitrogen oxides and oxyanions are generally slow. In practice, $N_2O$ has been found to be inert.

**(b)  Zinc metal is added to aqueous acidic sodium triiodide.** The overall reaction is:

$$Zn\ (s)\ +\ I_3^-\ (aq)\ \rightarrow\ Zn^{2+}\ (aq)\ +\ 3I^-\ (aq)$$

Since $E°$ values for the $Zn^{2+}/Zn$ and $I_3^-/I^-$ couples are −0.76 V and 0.54 V, respectively (these potentials are given in **Appendix 4**), $E°$ for the net reaction above is 0.54 V + 0.76 V = 1.30 V. Since the net potential is positive, this *is* a favorable reaction, and it should be kinetically facile if the zinc metal is finely divided and thus well exposed to the solution.

**(c)  $I_2$ is added to excess aqueous acidic $HClO_3$.** Since $E°$ values for the $I_2/I^-$ and $ClO_3^-/ClO_4^-$ couples are 0.54 V and 1.20 V, respectively (see **Appendix 6**), the net reaction involving the reduction of $I_2$ to $I^-$ and the oxidation of $ClO_3^-$ to $ClO_4^-$ will have a net negative potential, $E° = 0.54$ V + (−1.20 V) = −0.66 V. Therefore, this net reaction will not occur. However, $E°$ values for the $IO_3^-/I_2$ and $ClO_3^-/Cl^-$ couples are 0.54 V and 1.45 V, respectively, the following net reaction will occur, with a net $E° = 0.91$ V:

half-reaction:  $ClO_3^-\ (aq)\ +\ 6H^+\ +\ 6e^-\ \rightarrow\ Cl^-\ (aq)\ +\ 3H_2O\ (l)$

half-reaction:  $I_2\ (s)\ +\ 6H_2O\ \rightarrow\ 2IO_3^-\ (aq)\ +\ 6H^+\ +\ 10e^-$

net reaction: $3I_2\ (s)\ +\ 5ClO_3^-\ (aq)\ +\ 3H_2O\ (l)\ \rightarrow\ 6IO_3^-\ (aq)\ +\ 5Cl^-\ (aq)\ +\ 6H^+\ (aq)$

**8.7    Will acid or base most favor the following half-reactions? (a) $Mn^{2+} \rightarrow$ $MnO_4^-$.** We can answer questions such as this by applying Le Chatelier's principle to the complete, balanced half-reaction, which in this case is

$$Mn^{2+}\ (aq)\ +\ 4H_2O\ (l)\ \rightarrow\ MnO_4^-\ (aq)\ +\ 8H^+\ (aq)\ +\ 5e^-$$

Since hydrogen ions are produced by this oxidation half-reaction, raising the pH will favor the reaction. (Alternatively, we could come to the same conclusion by using the Nernst equation.) Thus, this reaction is favored in basic solution. At a sufficiently high pH, $Mn^{2+}$ will precipitate as $Mn(OH)_2$. The rate of oxidation for this solid will be much slower than for dissolved species.

(b) $ClO_4^- \rightarrow ClO_3^-$. The balanced half-reaction is

$$ClO_4^-\ (aq)\ +\ 2H^+\ (aq)\ +\ 2e^-\ \rightarrow\ ClO_3^-\ (aq)\ +\ H_2O\ (l)$$

Since hydrogen ions are consumed by this reduction half-reaction, lowering the pH will favor the reaction. Thus, this reaction is favored in acidic solution. This is the reason that perchloric acid is a dangerous oxidizing agent, even though salts of the perchlorate ion are frequently stable in neutral or basic solution.

(c) $H_2O_2 \rightarrow O_2$. The balanced half-reaction is

$$H_2O_2\ (aq)\ \rightarrow\ O_2\ (g)\ +\ 2H^+\ (aq)\ +\ 2e^-$$

Since hydrogen ions are produced by this oxidation half-reaction, raising the pH will favor the reaction. Thus, the reaction is favored in basic solution. This means that hydrogen peroxide is a better reducing agent in base than in acid.

(d) $I_2 \rightarrow I^-$. Since protons are neither consumed nor produced in this reduction half-reaction, and since $I^-$ is not protonated in aqueous solution (because HI is a very strong acid), this reaction has the same potential in acidic or basic solution, 0.535 V (this can be confirmed by consulting **Appendix 4**).

**8.8     Comment on the mechanisms of the following reactions: (a) HOI ($aq$) + I$^-$ ($aq$) $\rightarrow$ I$_2$ ($s$) + OH$^-$ ($aq$).** The transformation of reactants into products can be envisioned to take place by the simple transfer of an "I$^+$" ion from

HOI to I⁻, as shown below.  Therefore, this is probably an atom transfer reaction.

$$HOI + I^- \rightarrow HOI\cdots I^- \rightarrow HO^-\cdots I^+\cdots I^- \rightarrow HO^-\cdots I_2 \rightarrow HO^- + I_2$$

(b)  $[Co(phen)_3]^{3+}$ (aq) + $[Cr(bipy)_3]^{2+}$ (aq) → $[Co(phen)_3]^{2+}$ (aq) + $[Cr(bipy)_3]^{3+}$ (aq).  This reaction takes place without any net change in the coordination spheres of the two metal ions, since $Co^{3+}$ and $Cr^{3+}$ are kinetically inert to substitution (see **Section 7.8**).  The cobalt ion has three phenanthroline ligands whether it is +3 or +2 and the chromium ion has three bipyridine ligands whether it is +2 or +3.  Therefore, this is probably a simple outer-sphere electron transfer reaction.

(c)  $IO_3^-$ (aq) + 8I⁻ (aq) + 6H⁺ (aq) → $3I_3^-$ (aq) + $3H_2O$ (l).  Each of the $I_3^-$ ions are formed by the reaction of I⁻ with $I_2$ (see the solution to **End-of-Chapter Exercise 8.6(b)**).  It is unlikely that $I_2$ could be formed by a simple single step reaction of I⁻ with $IO_3^-$, since $IO_3^-$ must lose all three of its oxygen atoms.  Therefore, this is probably a multistep mechanism.

**8.9     Determine the standard potential for the reduction of $ClO_4^-$ to $Cl_2$.**
The Latimer diagram for chlorine in acidic solution is given in **Appendix 6** and the relevant portion of it is reproduced below:

$$ClO_4^- \xrightarrow{1.201\ V} ClO_3^- \xrightarrow{1.181\ V} HClO_2 \xrightarrow{1.674\ V} HClO \xrightarrow{1.630\ V} Cl_2$$

$$+7 \qquad\qquad +5 \qquad\qquad +3 \qquad\qquad +1 \qquad\qquad 0$$

To determine the potential for any couple, we must calculate the *weighted average* of the potentials of intervening couples.  In general terms it is

$$(n_1E^{\circ}_1 + n_2E^{\circ}_2 + ... + n_nE^{\circ}_n)/(n_1 + n_2 + ... + n_n)$$

and in this specific case it is

$$((2)(1.201\ V) + (2)(1.181\ V) + (2)(1.674) + (1)(1.630\ V))/(2 + 2 + 2 + 1) = 1.392\ V$$

Thus, the standard potential for the $ClO_4^-$, $Cl_2$ couple is 1.399 V. The balanced half-reaction for this reduction is

$$2ClO_4^- \,(aq) + 16H^+ \,(aq) + 14e^- \rightarrow Cl_2 \,(g) + 8H_2O \,(l)$$

Note that the point for $ClO_4^-$ in the Frost diagram shown in **Figure 8.15** has a $y$ value of 9.80, which is $nE°/V$ or 7(1.392).

**8.10    Convert reduction potentials into a series of vectors.** These are all one-electron processes, so the slopes of the three vectors are simply the reduction potentials for $[Fe(CN)_6]^{3-} \rightarrow [Fe(CN)_6]^{4-}$ (0.36 V), $Fe^{3+} \rightarrow Fe^{2+}$ (0.77 V), and $Cl_2 \rightarrow Cl^-$ (1.36 V). The vectors are shown in the diagram below, which also includes the vector for the $H^+$, $H_2$ couple as a reference.

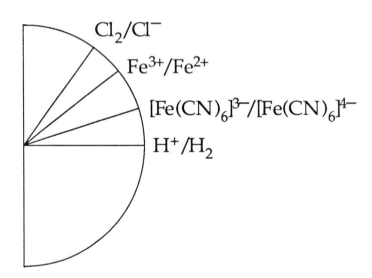

**8.11 Find the approximate potential of an aerated lake at pH = 6, and predict the predominant species for: (a) Fe.** According to **Figure 8.9**, the potential range for surface water at pH = 6 is 0.5 – 0.6 V, so a value of 0.55 V can be used as the approximate potential of an aerated lake at this pH. Inspection of the Pourbaix diagram for iron (**Figure 8.9**) shows that at pH - 6 and $E$ = 0.55 V, the stable species of iron is $Fe_2O_3$. Therefore, this compound of iron would predominate.

**(b) Mn.** Inspection of the Pourbaix diagram for manganese (**Figure 8.10**) shows that at pH = 6 and $E$ = 0.55 V, the stable species of manganese is $Mn_2O_3$. Therefore, this compound of manganese would predominate.

**(c) S.** At pH = 0, the potential for the $HSO_4^{2-}/S$ couple is 0.387 V (this value was calculated using the weighted average of the potentials given in the Latimer diagram for sulfur in **Appendix 4**), so the lake will oxidize $S_8$ all the way to $SO_4^{2-}$. At pH = 14, the potentials for intervening couples are all negative, so $SO_4^{2-}$ would again predominate. Therefore, $SO_4^{2-}$ is the predominant sulfur species at pH = 6.

**(d) Cl.** The predominant species at pH = 6 and $E$ = 0.55 V is $Cl^-$.

**8.12 What is the maximum $E$ for an anaerobic environment rich in $Fe^{2+}$ and $H_2S$?** Any species capable of oxidizing either $Fe^{2+}$ or $H_2S$ at pH = 6 cannot survive in this environment. According to the Pourbaix diagram for iron (**Figure 8.9**), the potential for the $Fe_2O_3$, $Fe^{2+}$ couple at pH = 6 is approximately 0.3 V. Using the Latimer diagrams for sulfur in acid and base (see **Appendix 4**), the $H_2S$, S potential at pH = 6 can be calculated as follows:

$$0.14 \text{ V} - (6/14)(0.14 \text{ V} - (-0.45 \text{ V})) = -0.11 \text{ V}$$

Any potential higher than this will oxidize hydrogen sulfide to elemental sulfur. Therefore, as long as $H_2S$ is present, the maximum potential possible is approximately –0.1 V.

**8.13 How will edta$^{4-}$ complexation affect $M^{2+} \rightarrow M^0$ reductions?** Since edta$^{4-}$ forms very stable complexes with $M^{2+}$ (*aq*) ions of Period 4 *d*-block elements but *not* with the zerovalent metal atoms, the reduction of a $M(edta)^{2-}$ complex will be more difficult than the reduction of the analogous $M^{2+}$ aqua ion. Since the reductions are more difficult, the reduction potentials become less positive (or more negative, as the case may be). This is represented in the equation below. The reduction of the $M(edta)^{2-}$ complex includes a decomplexation step, with a positive free energy change. The reduction of $M^{2+}$ (*aq*) does not require this additional expenditure of free energy.

$$M(edta)^{2-} \xleftarrow{\quad -edta^{4-} \quad} M^{2+} \xrightarrow{\quad +2\,e^- \quad} M^0$$

A drawing of a metal complex of $edta^{4-}$ is shown with the answer to **End-of-Chapter Exercise 7.21**.

**8.14    For which *d*-block elements is the group oxidation number reached by a stable species?** Frost diagrams for the first series of *d*-block elements in acidic solution are shown in **Figure 8.11**. The elements scandium-manganese can achieve the group oxidation number, but the chromium species, $Cr_2O_7^{2-}$, and the manganese species, $MnO_4^-$, are very strong oxidants. The elements Fe-Zn do not form any species with the group oxidation number. These results are summarized in the diagram below:

| Sc | Ti | V | Cr | Mn | Fe | Co | Ni | Cu | Zn |
|----|----|---|----|----|----|----|----|----|----|
| s  | s  | s | o  | o  | n  | n  | n  | n  | n  |

s = group oxidation number is a stable species

o = group oxidation number is a strongly oxidizing species

n = group oxidation number is not reached by a stable species

At this point in your studies, you should know the Period 4 $d$-block metals well. Try to integrate your chemical knowledge with the positions of the elements in the periodic table.

**8.15 Trends in the stability of the group oxidation number on descending a group.** On descending a group in the $d$-block, the group oxidation number becomes more and more stable. Thus, W(VI) is more stable than Mo(VI), which is more stable than Cr(VI). This can be seen by comparing reduction potentials for the following half-reactions in acid solution (data from **Appendix 4**):

$$Cr_2O_7^{2-} (aq) + 14H^+ (aq) + 12e^- \rightarrow 2Cr^0 + 7H_2O\ (l) \qquad E^\circ = 0.64\ V$$

$$H_2MoO_4 (aq) + 6H^+ (aq) + 6e^- \rightarrow Mo^0 + 4H_2O\ (l) \qquad E^\circ = 0.114\ V$$

$$WO_3 (s) + 6H^+ (aq) + 6e^- \rightarrow W^0 + 3H_2O\ (l) \qquad E^\circ = -0.090\ V$$

In the $p$-block, the group oxidation number is more stable for a Period 3 element compared with its heavier congeners:

$$Al^{3+} (aq) + 3e^- \rightarrow Al\ (s) \qquad E^\circ = -1.68\ V$$

$$Tl^{3+} (aq) + 3e^- \rightarrow Tl\ (s) \qquad E^\circ = 0.72\ V$$

**8.16 The most stable oxidation states of the actinides.** From the Frost diagrams in **Figure 8.14**, the most stable species for the first few actinides in acidic aqueous solution are $Ac^{3+}$, $Th^{4+}$, $PaO_2^+$ (i.e. Pa(V)), and $U^{4+}$. Thorium can be reduced from $Th^{4+}$ to $Th^{3+}$ only with great difficulty, and even $U^{4+}$ is difficult to reduce to $U^{3+}$. From neptunium on, the most stable species is the trivalent cation. In constrast, the most stable species in acidic aqueous solution is the trivalent cation for *all* of the lanthanide elements. Although one of the earliest lanthanides, cerium, can exist as $Ce^{4+}$ in aqueous acid, it is a strong oxidant and so is readily reduced back to $Ce^{3+}$. Therefore, the actinides, at least the early ones, have a richer variety of oxidation states than do the lanthanides.

# CHAPTER 9

# HYDROGEN AND ITS COMPOUNDS

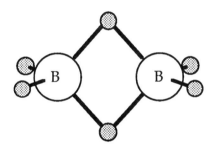

Diborane, $B_2H_6$, is the simplest member of a large class of compounds, the electron-deficient boron hydrides. Like all boron hydrides, it has a positive standard free energy of formation, and so cannot be prepared directly from boron and hydrogen. The bridge B–H bonds are longer and weaker than the terminal B–H bonds (1.32 vs. 1.19 Å).

## SOLUTIONS TO IN-CHAPTER EXERCISES

**9.1   Identifying an IR band by isotopic substitution.** Assuming that the Co–H and Co–D stretching force constants are the same, the vibrational frequencies are related to the reduced masses as follows:

$$\nu_{Co-D}/\nu_{Co-H} = (\mu_{Co-H}/\mu_{Co-D})^{1/2}$$

In many cases we can use the approximation:

$$(\mu_{Co-H}/\mu_{Co-D})^{1/2} \approx (m_H/m_D)^{1/2} = (1/2)^{1/2} = 0.7071$$

Let's see how good an approximation this is. Using $m_H = 1$, $m_D = 2$, and $m_{Co}$ = 59, $\mu_{Co-H} = (1/m_H) + (1/m_{Co}) = (1/1) + (1/59) = 1.0169$ and $\mu_{Co-D} = (1/m_D) + (1/m_{Co}) = (1/2) + (1/59) = 0.5169$, so

$$(\mu_{Co-H}/\mu_{Co-D})^{1/2} = (1.0169/0.5169)^{1/2} = (1.9673)^{1/2} = 0.7013$$

The error introduced by using 0.7071 instead of 0.7013 is less than 1%. Therefore, given that $v_{Co-H} = 1840$ cm$^{-1}$, $v_{Co-D}$ should be $(1840$ cm$^{-1})(0.7071)$ = 1301 cm$^{-1}$.

**9.2    Explain how a foreign radical can slow down a radical chain reaction.** This can happen if the added radical, X·, reacts with the product faster than with any of the reactants. In the case of $H_2 + Cl_2 \rightarrow 2HCl$, if X· + HCl $\rightarrow$ H· + XCl is a fast step, then the following steps would reform *some* starting material, slowing down the overall rate of HCl production:

$$H\cdot + H\cdot \rightarrow H_2 \qquad\qquad \cdot Cl + XCl \rightarrow Cl_2 + X\cdot$$

**9.3    Examples of hydrogen compounds.** Let us choose examples from Period 4: $CaH_2$ (Group 2), $CuH_x$ (Group 11, $0 < x \leq 1$, see **Figure 9.5**), and $H_2Se$ (Group 16/VI). Calcium hydride is a saline hydride and, like most salts, is a solid. The compound $CuH_x$, which is also a solid, is a metallic hydride. Neither of these two compounds can participate in hydrogen bonding. Hydrogen selenide is a molecular hydride. Since the electronegativity of Se is not as great as N, O, or F, $H_2Se$ does not exhibit hydrogen bonding and is a gas (not a liquid like $H_2O$).

**9.4    A likely mechanism for the reaction of Et$_3$PbH with CH$_3$Br.** The overall reaction is

$$Et_3PbH + CH_3Br \rightarrow Et_3PbBr + CH_4$$

Since the Pb–H bond is not very polar (the electronegativities of Pb and H are 2.33 and 2.20, respectively (**Table 1.9**)) and is very weak (from **Figure 9.9** we can extrapolate that the Pb–H bond enthalpy is probably $220 \pm 20$ kJ mol$^{-1}$), it is likely that the mechanism for this reaction involves the homolytic cleavage of the Pb–H bond and the formation of radical species such as $Et_3Pb\cdot$ and $\cdot CH_3$.

**9.5    Write equations for the reaction of $B_2H_6$ with propylene.** Since the H atoms in boron hydrides such as $B_2H_6$ are slightly hydridic, the reaction involves addition across the C=C double bond by hydride transfer:

$$B_2H_6 + 2THF \rightarrow 2BH_3 \cdot THF$$

$$BH_3 \cdot THF + CH_3CH=CH_2 \rightarrow CH_3CH_2CH_2BH_2 \cdot THF$$

Note the anti-Markovnikov addition of the >B–H moiety across the double bond.  Consult an organic chemistry text for the differences between Markovnikov and anti-Markovnikov additions.

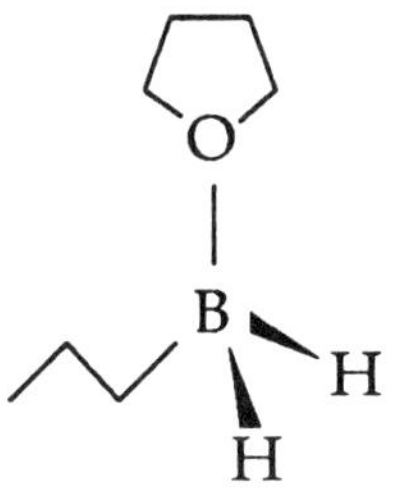

The THF complex of an organoborane

## SOLUTIONS TO END-OF-CHAPTER EXERCISES

**9.1    Properties of hydrides of the elements.    (a)  Position in the Periodic Table.** See **Figure 9.3**.

   **(b) Trends in $\Delta G_f^\circ$.** See **Table 9.5**.

   **(c)  Different molecular hydrides.**  Molecular hydrides are found in Groups 13/III through 17/VII.  Those in Group 13/III are electron-deficient, those in Group 14/IV are electron-precise, and those in Group 15/V through 17/VII are electron-rich.

**9.2    Name and classify the following: (a) $BaH_2$.** This compound is named barium hydride.  It is a saline hydride.

(b)  **$SiH_4$.** This compound is named silane.  It is an electron-precise molecular hydride.

(c)  **$NH_3$.** This familiar compound is known by its common name, ammonia, rather than by the systematic names azane or nitrane.  Ammonia is an electron-rich molecular hydride.

(d)  **$AsH_3$.** This compound is generally known by its common name, arsine, rather than by its systematic name, arsane.  It is also an electron-rich molecular hydride.

(e)  **$PdH_{0.9}$.** This compound is named palladium hydride.  It is a metallic hydride.

(f)  **HI.** This compound is known by its common name, hydrogen iodide, rather than by its systematic name, iodane.  It is an electron-rich molecular hydride.

**9.3    Chemical characteristics of hydrides. (a) Hydridic character.** Barium hydride is a good example, since it reacts with proton sources such as $H_2O$ to form $H_2$:

$$BaH_2\ (s)\ +\ 2H_2O\ (l)\ \rightarrow\ 2H_2\ (g)\ +\ Ba(OH)_2\ (s)$$

$$\text{net reaction:}\quad 2H^-\ +\ 2H^+\ \rightarrow\ 2H_2$$

(b)  **Brønsted acidity.** Hydrogen iodide is a good example, since it transfers its proton to a variety of bases, including pyridine (:py):

$$HI\ (g)\ +\ :py\ (g)\ \rightarrow\ [H{:}py]^+[I]^-\ (s)$$

(c)  **Lewis basicity.** Ammonia is a good example, since it forms acid-base complexes with a variety of Lewis acids, including $BF_3$:

$$NH_3\ (g)\ +\ BF_3\ (g)\ \rightarrow\ F_3BNH_3\ (s)$$

**9.4      Phases of hydrides of the elements.** Of the compounds listed in **End-of-Chapter Exercise 9.2**, $BaH_2$ and $PdH_{0.9}$ are solids, none is a liquid, and $SiH_4$, $NH_3$, $AsH_3$, and $HI$ are gases (see **Figure 9.7**). Only $PdH_{0.9}$ is likely to be a good electrical conductor.

**9.5      The structures of $H_2Se$, $P_2H_4$, and $H_3O^+$.** The Lewis structures of these three species are:

$$H-\ddot{\underset{..}{Se}}-H \qquad H-\underset{\underset{H}{|}}{\overset{\overset{H}{|}}{\underset{..}{P}}}-\underset{\underset{..}{|}}{\overset{\overset{H}{|}}{P}}-H \qquad \left[H-\underset{..}{\overset{\overset{H}{|}}{O}}-H\right]^+$$

According to VSEPR theory (**Section 2.2**), $H_2Se$ should be bent, and so it belongs to the $C_{2v}$ point group; $H_3O^+$ should be trigonal pyramidal (like $NH_3$), and so it belongs to the $C_{3v}$ point group; each phosphorus atom of $P_2H_4$ should have local pyramidal structure — if the molecule adopts the skew conformation (see the Newman diagram below), then it belongs to the $C_2$ point group (the $C_2$ axis bisects the P–P bond).

A drawing of the structure of $P_2H_4$. The P–P and P–H bond distances are 2.22 and 1.42 Å, respectively, and the P–P–H bond angles are all about 94° (cf. $PH_3$, in which the H–P–H bond angles are 93.8° (see **Table 9.4**)).

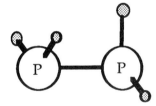

A Newman projection of the skew, or gauche, conformation of $P_2H_4$. The only element of symmetry that this structure possesses is a $C_2$ axis that bisects the P–P bond. Therefore, it has $C_2$ symmetry.

**9.6    The reaction that will give the highest proportion of HD.** Reactions **(a)** and **(c)** both involve the production of both H and D atoms on the surface of a metal. The recombination of these atoms will give a statistical distribution of $H_2$ (25%), HD (50%), and $D_2$ (25%). However, reaction **(b)** involves a source of protons that is 100% $^2H^+$ (or $D^+$) and a source of hydride ions that is 100% $^1H^-$:

$$D_2O\ (l)\ +\ NaH\ (s)\ \rightarrow\ HD\ (g)\ +\ NaOD\ (s)$$

$$\text{net reaction:}\quad D^+\ +\ H^-\ \rightarrow\ HD$$

Thus, reaction **(b)** will produce 100% HD and no $H_2$ or $D_2$.


**9.7    NMR spectra of $PH_3$ and $P(OCH_3)_3$.** For both compounds, there is a single type of $^1H$ atom coupled to one $^{31}P$ atom, so the $^1H$ NMR spectrum of each compound will be a doublet. Since the P–H coupling constant is probably greater in $PH_3$ than in $P(OCH_3)_3$ (fewer bonds between the two atoms), the splitting in the $^1H$ NMR spectrum of $PH_3$ will be larger than in the spectrum of $P(OCH_3)_3$. These spectra are shown below as schematics (i.e. the peaks are represented as straight lines with no apparent linewidth — real NMR peaks have a Lorentzian shape with a finite linewidth)

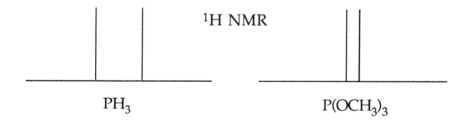

For both compounds, there is a single type of $^{31}P$ atom coupled to several $^1H$ atoms, so the $^{31}P$ NMR spectrum of each compound will consist of one multiplet. Since the multiplet for a nucleus coupled to $n$ equivalent $I = 1/2$ nuclei always consists of $n + 1$ peaks, the $^{31}P$ NMR spectrum of $PH_3$ will be a 1:3:3:1 quartet while that of $P(OCH_3)_3$ will be a ten line 1:9:36:84:126:126:84:36:9:1 pattern (use Pascal's triangle for $n = 3$ and $n = 9$), as shown below.

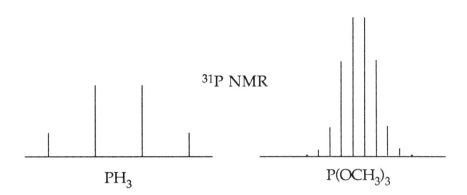

9.8    **Balanced chemical equations for industrial preparations of $H_2$.** The three major industrial methods of production of $H_2$ are (i) steam reforming of hydrocarbons, principally methane, (ii) the water-gas reaction, and (iii) the dehydrogenation of alkanes.  Balanced equations are:

(i)    $CH_4\ (g)\ +\ H_2O\ (g)\ \rightarrow\ 3H_2\ (g)\ +\ CO\ (g)$

(ii)    $C\ (s)\ +\ H_2O\ (g)\ \rightarrow\ H_2\ (g)\ +\ CO\ (g)$

(iii)    e.g. $C_2H_6\ (g)\ \rightarrow\ C_2H_4\ (g)\ +\ H_2\ (g)$

In the laboratory, more convenient methods are (i) the electrolysis of water, (ii) the reaction of metallic zinc with hydrochloric acid, and (iii) the hydrolysis of saline hydrides.  Balanced equations for these reactions are:

(i)    $2H_2O\ (l)\ +\ electricity\ \rightarrow\ 2H_2\ (g)\ +\ O_2\ (g)$

(ii)    $Zn\ (s)\ +\ 2HCl\ (aq)\ \rightarrow\ H_2\ (g)\ +\ ZnCl_2\ (aq)$

(iii)    $NaH\ (s)\ +\ H_2O\ (l)\ \rightarrow\ H_2\ (g)\ +\ NaOH\ (aq)$

9.9    **Most likely to undergo radical reactions.**  Of the compounds $H_2O$, $NH_3$, $(CH_3)_3SiH$, and $(CH_3)_3SnH$, the tin compound is the most likely to undergo radical reactions with alkyl halides.  This is because the Sn–H bond in $(CH_3)_3SnH$ is less polar *and* weaker than either O–H, N–H, or Si–H bonds. The formation of radicals involves the homolytic cleavage of the bond between the central element and hydrogen, and a weak nonpolar bond undergoes homolysis most readily.

**9.10    Arrange $H_2O$, $H_2S$, and $H_2Se$ in order of: (a) Increasing acidity.** As discussed in **Section 5.3**, acidities of $EH_n$ increase down a group in the *p*-block, mostly because the decrease in E–H bond enthalpy lowers the proton affinity of $EH_{n-1}^-$ (E is a generic *p*-block element).  Therefore, the order of increasing acidity is $H_2O < H_2S < H_2Se$.

**(b)  Increasing basicity toward a hard acid.**  In general, soft character increases down a group, so the hardest base of these three compounds is $H_2O$. The order of increasing basicity toward a hard acid is $H_2Se < H_2S < H_2O$.

**9.11    The synthesis of binary hydrogen compounds.**  The three main methods of synthesis of binary hydrogen compounds are (i) direct combination of the elements, (ii) protonation of a Brønsted base, and (iii) metathesis using a compound such as LiH, $NaBH_4$, or $LiAlH_4$.  The first method is limited to those binary hydrogen compounds that are exoergic. An example is:

$$(i)\quad 2Li\ (s)\ +\ H_2\ (g)\ \rightarrow\ 2LiH\ (s)$$

The second method can be used for the preparation of $EH_n$ compounds when a source of the $E^{n-}$ anion is available.  An example is:

$$(ii)\quad CaF_2\ (s)\ +\ H_2SO_4\ (l)\ \rightarrow\ 2HF\ (g)\ +\ CaSO_4\ (s)$$

Almost all of the hydrogen fluoride that is prepared industrially is made this way.  The third method can be used to convert the chlorides of many elements E to the corresponding hydrides, as in the following example:

$$PCl_3\ (l)\ +\ 3LiH\ (s)\ \rightarrow\ PH_3\ (g)\ +\ 3LiCl\ (s)$$

**9.12    Give laboratory methods of synthesizing: (a) $H_2Se$.** Since hydrogen selenide is endoergic (see **Table 9.5**), it cannot be prepared from elemental hydrogen and selenium. It can, however, be prepared by protonating a salt of the $Se^{2-}$ ion, as in the following equations:

$$2Na\ (s)\ +\ Se\ (s)\ \rightarrow\ Na_2Se\ (s)$$

$$Na_2Se\ (s)\ +\ 2H_3PO_4\ (l)\ \rightarrow\ H_2Se\ (g)\ +\ 2NaH_2PO_4\ (s)$$

**(b)  $SiD_4$.** Both $SiCl_4$ and $LiAlH_4$ are exoergic compounds that can be prepared from their constituent elements. When reacted together, they form $SiH_4$. Therefore, the following reaction scheme can be used to prepare $SiD_4$:

$$2Li\ (s)\ +\ D_2\ (g)\ \rightarrow\ 2LiD\ (s)\qquad 2Al\ (s)\ +\ 3Cl_2\ (g)\ \rightarrow\ 2AlCl_3\ (s)$$

$$AlCl_3\ (s)\ +\ 4LiD\ (s)\ \rightarrow\ LiAlD_4\ (s)\ +\ 3LiCl\ (s)\qquad Si\ (s)\ +\ 2Cl_2\ (g)\ \rightarrow\ SiCl_4\ (l)$$

$$SiCl_4\ (l)\ +\ LiAlD_4\ (s)\ \rightarrow\ LiAlCl_4\ (s)\ +\ SiD_4\ (g)$$

**(c)  $Ge(CH_3)_2H_2$.** Once again, $LiAlH_4$ can be used to accomplish a H/Cl metathesis, as in the following reaction:

$$2Ge(CH_3)_3Cl_2\ (l)\ +\ LiAlH_4\ (s)\ \rightarrow\ 2Ge(CH_3)_3H_2\ (l)\ +\ LiAlCl_4\ (s)$$

**(d)  $SiH_4$ from Si and HCl.** This reaction involves the oxidation of Si by HCl. To balance this redox process, something must be reduced, and the most likely candidate is $H^+$:

$$Si\ (s)\ +\ 3HCl\ (g)\ \rightarrow\ SiHCl_3\ (l)\ +\ H_2\ (g)$$

The trichlorosilane is then heated, whereupon it undergoes a redistribution reaction, as shown below:

$$4SiHCl_3\ (l)\ +\ heat\ \rightarrow\ SiH_4\ (g)\ +\ 3SiCl_4\ (l)$$

**9.13    Is $B_2H_6$ stable in air?** No, this compound reacts so vigorously with air that it spontaneously inflames when exposed to air (compounds that display this behavior are called pyrophoric). It can react with both oxygen and moisture in the air, according to the following reactions:

$$B_2H_6 \ (g) \ + \ 3O_2 \ (g) \ \rightarrow \ B_2O_3 \ (s) \ + \ 3H_2O \ (l) \quad (or \ 2B(OH)_3 \ (s))$$

$$B_2H_6 \ (g) \ + \ 3H_2O \ (l) \ \rightarrow \ 2B(OH)_3 \ (s) \ + \ 3H_2 \ (g)$$

In order to transfer diborane from a storage bulb, in which it exerts a pressure of 200 Torr, to a reaction vessel containing diethyl ether, we must first cool the reaction vessel with liquid nitrogen to –196 °C (77 K). This will freeze the ether. Then we remove all of the noncondensable gases (e.g. $N_2$) from the reaction vessel with the vacuum system. Finally, with the reaction vessel still cooled to –196 °C, we disconnect the vacuum system from the reaction vessel and connect the storage bulb to the reaction vessel. At this temperature, the sample of diborane will completely condense in the reaction vessel. Once we seal the vessel (by closing its valve), we can allow it to warm up to room temperature, at which point we will have a *solution* of diborane in diethyl ether.

**9.14   Compare $BH_4^-$, $AlH_4^-$, and $GaH_4^-$.** Since Al has the lowest electronegativity of the three elements B (2.04), Al (1.61), and Ga (1.81, see **Table 1.9**), the Al–H bonds of $AlH_4^-$ are more hydridic than the B–H bonds of $BH_4^-$ or the Ga–H bonds of $GaH_4^-$. Therefore, since $AlH_4^-$ is more "hydride-like," it is the strongest reducing agent. The reaction of $GaH_4^-$ with aqueous HCl is as follows:

$$GaH_4^- \ (aq) \ + \ 4HCl \ (aq) \ \rightarrow \ GaCl_4^- \ (aq) \ + \ 4H_2 \ (g)$$

**9.15   The formation of pure Si from crude Si.** Crude silicon is treated with gaseous HCl (not aqueous HCl) to form $SiHCl_3$, which undergoes a redistribution reaction at moderately high temperature to form $SiH_4$ and $SiCl_4$. These two compounds are separated by fractional distillation (their normal boiling points are –112 °C and 58 °C, respectively), and the highly purified $SiH_4$ is decomposed at 500 °C to pure Si and $H_2$. Balanced equations for this process are shown below:

$$Si\ (s)\ +\ 3HCl\ (g)\ \rightarrow\ SiHCl_3\ (l)\ +\ H_2\ (g)$$

$$4SiHCl_3\ (l)\ +\ heat\ \rightarrow\ SiH_4\ (g)\ +\ 3SiCl_4\ (l)$$

$$500\ °C:\quad SiH_4\ (g)\ \rightarrow\ Si\ (s)\ +\ 2H_2\ (g)$$

**9.16 Compare Period 2 and Period 3 hydrogen compounds.** One important difference between Period 2 and Period 3 hydrogen compounds is their relative stabilities. The Period 2 compounds, except for $B_2H_6$, are all exoergic (see **Table 9.5**). Their Period 3 homologues are either much less exoergic or are endoergic (cf. HF and HCl, for which $\Delta G_f° = -273.2$ and $-95.3$ kJ mol$^{-1}$, and $NH_3$ and $PH_3$, for which $\Delta G_f° = -16.5$ and $+13.4$ kJ mol$^{-1}$). Another important difference is that Period 2 compounds tend to be weaker Brønsted acids and stronger Brønsted bases than their Period 3 homologues. The bond angles in Period 2 hydrogen compounds reflect a greater degree of $sp^3$ hybridization than the homologous Period 3 compounds (cf. the H–O–H and H–N–H bond angles of water and ammonia, which are 104.5° and 106.6°, respectively, to the H–S–H and H–P–H bond angles of hydrogen sulfide and phosphine, which are 92° and 93.8°, respectively). Finally, several Period 2 compounds exhibit strong hydrogen bonding, namely HF, $H_2O$, and $NH_3$, while their Period 3 homologues do not (see **Figure 9.7**).

**9.17 Describe the compound formed between water and Kr.** This compound is called a clathrate hydrate (see **Section 9.16**). It consists of cages of water molecules, all hydrogen bonded together, each surrounding a single krypton atom (cf. the structure of the clathrate hydrate of molecular chlorine, shown in **Figure 9.13**). Strong dipole-dipole forces hold the cages together, while weaker van der Waals forces hold the krypton atoms in the centers of their respective cages.

**9.18 Molecular orbitals for $HF_2^-$.** The three atoms of the bifluoride ion are colinear. Nevertheless, the 3-center bonding in this ion is somewhat analogous to the 3-center bonding of the bent B–H–B bridges in diborane, which was discussed in **Section 3.2** and the orbitals for which are shown in **Figure 3.10**. The major difference between bifluoride ion and B–H–B bridge bonds is that the former is an example of 3-center, 4-electron bonding while the latter is an example of 3-center, 2-electron bonding. In bifluoride ion, the bottom (bonding) and middle (nonbonding) molecular orbitals are filled while the top (antibonding) molecular orbital is empty.

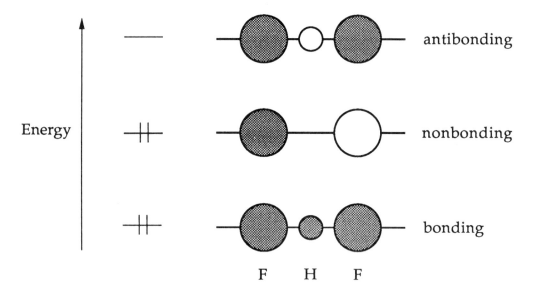

**9.19   Potential energy surfaces for hydrogen bonds.** There are two important differences between the potential energy surfaces for the hydrogen bond between $H_2O$ and $Cl^-$ ion and for the hydrogen bond in bifluoride ion, $HF_2^-$. The first difference is that the surface for the $H_2O$, $Cl^-$ system has a double minimum (as do most hydrogen bonds), since it is a relatively weak hydrogen bond, while the surface for the bifluoride ion has a single minimum (characteristic of only the strongest hydrogen bonds). The second difference is that the surface for the $H_2O$, $Cl^-$ system is not symmetric, since the proton is bonded to two different atoms (oxygen and chlorine), while the surface for bifluoride ion is symmetric. The two surfaces are shown below.

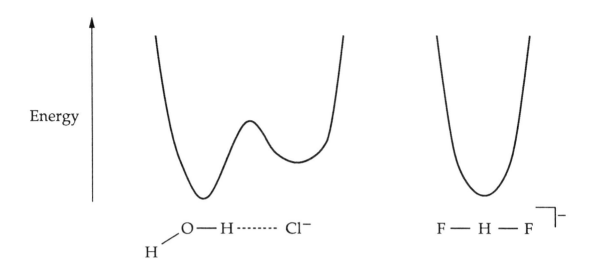

**9.20** **Describe the NMR spectra of PF₃ shown in Figure 9.18.** The $^{19}$F (a) and $^{31}$P (b) NMR spectra of PF₃ are shown below. The relative intensities of the peaks in each spectrum are given above each peak. In the $^{19}$F spectrum, one symmetrical doublet (i.e. a 1:1 doublet) is observed, since (i) all three fluorines are equivalent, (ii) $^{31}$P is 100% abundant, and (iii) $I = 1/2$ for $^{31}$P. In the $^{31}$P spectrum, one 1:3:3:1 quartet is observed, since (i) there is only one $^{31}$P atom in the compound, (ii) it is coupled to *three* equivalent $I = 1/2$ $^{19}$F atoms, and (iii) the $^{19}$F atoms are 100% abundant. The particular splitting pattern for a nucleus coupled to three equivalent $I = 1/2$ nuclei is given by Pascal's triangle, which is discussed in the **Further Information** section of **Chapter 9**.

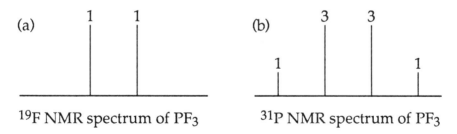

$^{19}$F NMR spectrum of PF₃          $^{31}$P NMR spectrum of PF₃

**9.21** **Sketch the $^{11}$B NMR spectra for HB(CH₃)₃⁻ and H₂B(CH₃)₂⁻.** Given that only the hydrogen atoms directly bonded to boron are spin coupled with it, the $^{11}$B spectrum of the first ion will show a 1:1 doublet, representing coupling to one $^1H$ atom, while the $^{11}B$ spectrum of the second ion will show a 1:2:1 triplet, representing coupling to two $^1H$ atoms.

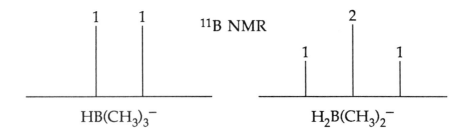

**9.22    Sketch the $^1H$ NMR spectrum for $BH_4^-$.** Given that naturally occuring boron consists of about 80% $^{11}B$ ($I = 3/2$) and about 20% $^{10}B$ ($I = 3$), there are really two types of $BH_4^-$ ions in the sample, $^{11}BH_4^-$ and $^{10}BH_4^-$. Each type of ion gives rise to a given multiplet pattern, a 1:1:1:1 quartet for $^{11}BH_4^-$ ($2I + 1 = 2(3/2) + 1 = 4$) and a 1:1:1:1:1:1:1 septet for $^{10}BH_4^-$ ($2 \times 3 + 1 = 7$). Since the electronic environment of the $^1H$ atoms are the same in the two ions, and since the four $^1H$ atoms in each ion are equivalent, the same chemical shift is observed for each $^1H$ atom in both types of ion (i.e. the centers of the two multiplet patterns are superimposed). It turns out that the coupling between the $^{11}B$ and $^1H$ atoms is greater, by about a factor of three, than the coupling between the $^{10}B$ and $^1H$ atoms, but this cannot be predicted from the information given. The two individual multiplet patterns and the expected $^1H$ NMR spectrum of the $BH_4^-$ ion are shown below.

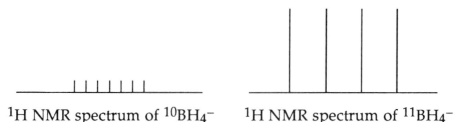

$^1H$ NMR spectrum of $^{10}BH_4^-$         $^1H$ NMR spectrum of $^{11}BH_4^-$

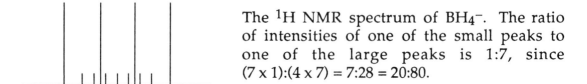

The $^1H$ NMR spectrum of $BH_4^-$. The ratio of intensities of one of the small peaks to one of the large peaks is 1:7, since $(7 \times 1):(4 \times 7) = 7:28 = 20:80$.

# CHAPTER 10

# MAIN GROUP ORGANOMETALLICS

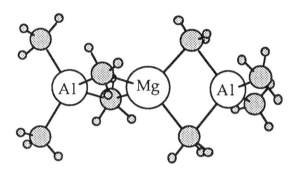

A drawing of $Mg[Al(CH_3)_4]_2$, an organometallic compound containing magnesium, aluminum, and methyl groups. The compound is essentially covalent and is soluble in nonpolar solvents such as cyclopentane and toluene.

---

## SOLUTIONS TO IN-CHAPTER EXERCISES

**10.1   Classifying and predicting M–C bond forming reactions.** In this case, we are given an electropositive metal, magnesium, and the organometallic compound of a less electropositive metal, dimethylmercury. A trans-metallation reaction will take place:

$$Mg\ (s)\ +\ Hg(CH_3)_2\ (l)\ \rightarrow\ Mg(CH_3)_2\ (s)\ +\ Hg\ (l)$$

**10.2   Predicting the products of thermal decomposition.** The Pb–C bonds in $Pb(CH_3)_4$ are weak and readily undergo homolysis in much the same way as $Bi(CH_3)_3$, discussed in the **Example**. β-Hydrogen elimination is not possible

145

with methyl groups, since there is no $\beta$ carbon atom. Therefore, the probable mode of thermal decomposition of tetramethyllead is:

$$Pb(CH_3)_4 \ (l) + heat \rightarrow Pb \ (s) + 2C_2H_6 \ (g)$$

**10.3    Propose a structure for Al$_2$($i$-Bu)$_4$H$_2$.** Since the tendency towards bridge structures is PR$_2^-$ > X$^-$ > H$^-$ > Ph$^-$ > R$^-$, the hydride ligands, and not the isobutyl ligands, will bridge the two aluminum atoms.

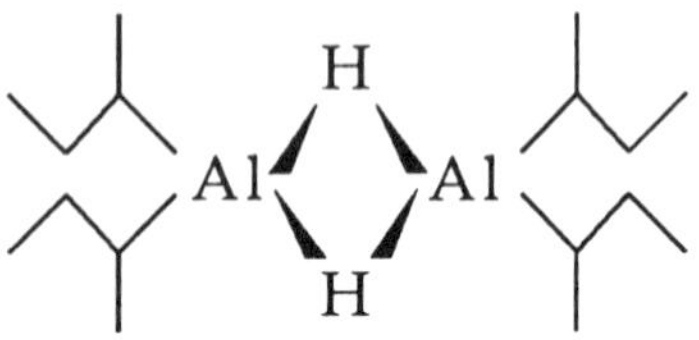

The structure of Al$_2$($i$-Bu)$_4$H$_2$

**10.4    The difference in structure between (H$_3$Si)$_3$N and (H$_3$C)$_3$N.** If the N atom lone pair of (H$_3$Si)$_3$N is delocalized onto the three Si atoms, it cannot exert its normal steric influence as predicted by VSEPR rules. Therefore, the N atom of (H$_3$Si)$_3$N is trigonal planar, whereas the N atom of (H$_3$C)$_3$N is trigonal pyramidal.

The structures of (H$_3$Si)$_3$N and (H$_3$C)$_3$N, excluding the hydrogen atoms.

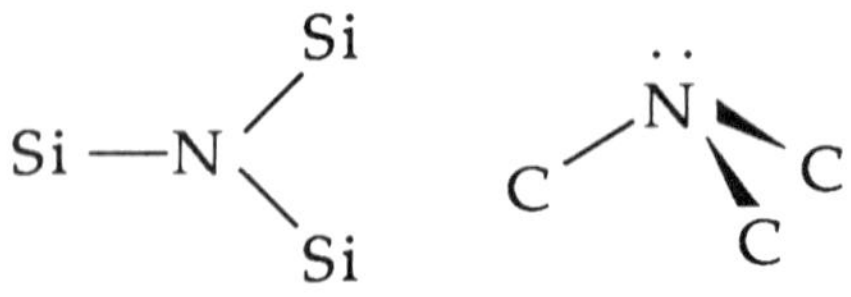

**10.5    The hydrolysis of Al$_2$(CH$_3$)$_6$ and Ga(CH$_3$)$_3$.** As mentioned at the beginning of **Section 10.11**, Ga–C bonds are more resistant to hydrolysis than Al–C bonds. Since Ga is more electronegative than Al, the CH$_3$ groups in Ga(CH$_3$)$_3$ have less carbanion character than the CH$_3$ groups in Al$_2$(CH$_3$)$_6$. Only one of the CH$_3$ groups in Ga(CH$_3$)$_3$ is cleaved by H$_2$O, as shown below:

$$Al_2(CH_3)_6 + 6H_2O \rightarrow 2Al(OH)_3 + 6CH_4$$

$$Ga(CH_3)_3 + nH_2O \rightarrow [(CH_3)_2GaOH]_n + nCH_4$$

**10.6 Compare the stable hydrogen compounds of Ge and As.** The trends exhibited by the stable hydrogen compounds of Ge and As follow the trends for their alkyl compounds, which were discussed in the **Example**. The compound $GeH_4$ is the only stable hydrogen compound of Ge; $GeH_2$ is unknown. This parallels the observation that only a few organo-germanium(II) compounds are known. In contrast, $AsH_3$ is the only stable hydrogen compound of As; $AsH_5$ is unknown. This parallels the fact that $As(CH_3)_5$ is unstable (one of the only pentavalent organoarsenic compounds is $AsPh_5$, which is mentioned in the text). The fact that the trends for hydrogen compounds and alkyl compounds are so similar should make sense to you, since hydrogen and carbon have similar electronegativities (2.20 vs. 2.55, respectively) and the strengths and polarities of element-hydrogen and element-carbon bonds are also quite similar.

## SOLUTIONS TO END-OF-CHAPTER EXERCISES

**10.1 Organometallic or not organometallic? (a) $B(CH_3)_3$.** According to the definition of the authors, a compound is organometallic if it contains at least one carbon-metal bond, and the suffix "metallic" includes the metalloids B, Si, and As. Therefore, since trimethylboron contains C–B bonds, it is an organometallic compound.

(b) $B(OCH_3)_3$. This compound also contains a boron atom and methyl groups, but it does not have any C–B bonds (only O–B, C–O, and C–H bonds). Therefore, trimethoxyboron (or trimethylborate) is not an organometallic compound.

(c) $(NaCH_3)_4$. The structure of this compound is similar to methyllithium (**Structural Drawing 2** in the text). Since it contains C–Na bonds, it is an organometallic compound.

(d) **$SiCl_3(CH_3)$.** This compound has three Si–Cl and one C–Si bonds around the tetrahedral central silicon atom. Since it contains at least one C–Si bond, methyltrichlorosilane is an organometallic compound.

(e) **$N(CH_3)_3$.** This compound, trimethylamine, does not even contain a metal or metalloid atom, so it cannot be an organometallic compound.

(f) **Sodium acetate.** This salt does not contain any C–Na bonds. The closest atoms to the $Na^+$ ions in the lattice would be the carboxylate oxygen atoms. It is not an organometallic compound.

(g) **$Na[B(C_6H_5)_4]$.** This salt also does not contain any C–Na bonds, but the $B(C_6H_5)_4^-$ anion contains four C–B bonds. Therefore, sodium tetraphenylborate is an organometallic compound. Specifically, it is the sodium salt of an organometallic anion.


**10.2   s- and p-block and Group 12 organometallics.** If you cannot do all of this exercise without consulting reference material, try to do as much as you can. Start by drawing a piece of the periodic table with the relevant elements drawn in (all of the elements from Groups 1, 2, 12, and 13, from Si on down in Group 14, and from As on down in Group 15). Then write out a formula for a representative methyl compound for each group, taking into account the normal oxidation state for elements in that group (+1 for Group 1, +2 for Groups 2 and 12, +3 for Groups 13 and 15, and +4 for Group 14). Then remember that saline methyl compounds, just like saline hydrides, are formed by the most electropositive elements, i.e. those of Groups 1 and 2. The saline methyl compounds are electron deficient (that is why their structures exhibit bridging methyl groups), as are the methyl compounds of Groups 12 and 13. Those of Group 14 are electron precise, i.e. the central atom has an octet with no lone pairs. The organometallic methyl compounds of Group 15 are electron rich. The central atom in $As(CH_3)_3$ and in similar compounds has an octet, but one of the pairs of electrons is a lone pair. A periodic table showing all of this information is shown below. The final part of this exercise is to recognize trends in $\Delta H_f^\circ$ in the p-block. Just as for the p-block hydrides (cf. **Tables 9.4 and 9.5**), the standard enthalpy of formation of p-block methyl compounds becomes less negative or more positive on going down a group, in large part because the C–M bond enthalpies become progressively weaker on going down a group (see **Tables 10.1** and **10.2**).

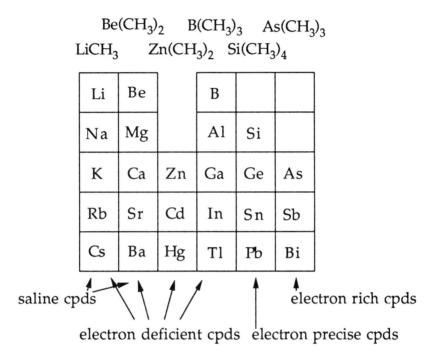

$$Be(CH_3)_2 \quad B(CH_3)_3 \quad As(CH_3)_3$$
$$LiCH_3 \quad Zn(CH_3)_2 \quad Si(CH_3)_4$$

**10.3   Formulas and alternative names. (a) Trimethylbismuth.** The formula is $Bi(CH_3)_3$. An alternative name is trimethylbismuthine, i.e. the methyl derivative of bismuthine, $BiH_3$.

**(b)   Tetraphenylsilane.** The formula is $Si(C_6H_5)_4$. An alternative name is tetraphenylsilicon(IV).

**(c)   Tetraphenylarsonium bromide.** The formula is $[As(C_6H_5)_4]Br$.

**(d)   Potassium tetraphenylborate.** The formula is $K[B(C_6H_5)_4]$. An alternative name is potassium tetraphenylboron(−1).

**10.4   Name, classify, and sketch each of the following compounds: (a) SiH(C$_2$H$_5$)$_3$.** This compound is named triethylsilane. It is an electron-precise organometallic compound. Its structure, containing a tetrahedral Si atom, is shown below.

(b) **BCl(C6H5)2.** This compound is named diphenylchloroborane. It is an electron deficient organometallic compound. Its structure, with trigonal planar bonding to the B atom, is also shown below.

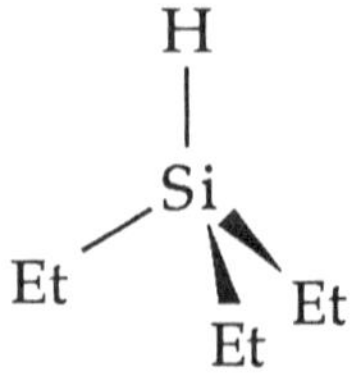

SiH(C₂H₅)₃        BCl(C₆H₅)₂                    Al₂Cl₂(C₆H₅)₄

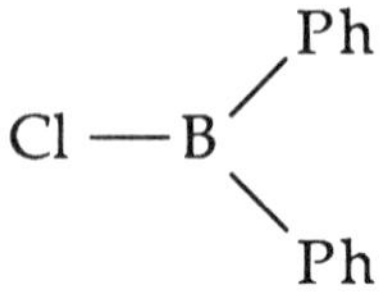

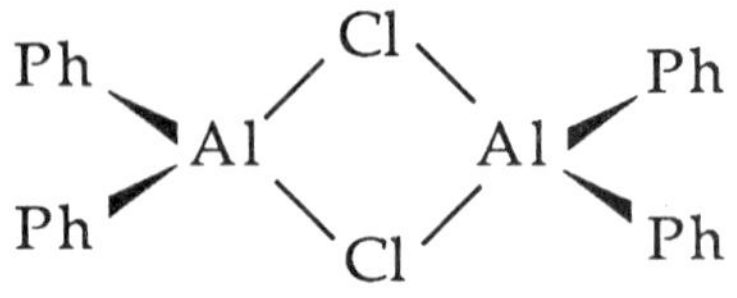

(c) **Al₂Cl₂(C6H5)4.** This compound is named tetraphenyldichlorodialuminum. It is a dimer of the electron-deficient monomer diphenylchloroaluminum. Its structure, a dimer with Cl atom bridges and tetrahedral Al atoms, is shown above.

(d) **Li4(C2H5)4.** This compound is named ethyllithium (or tetraethyltetralithium). It is a tetramer of the electron-deficient monomer ethyllithium. Its structure is shown at the right (one of the ethyl groups has been omitted for clarity).

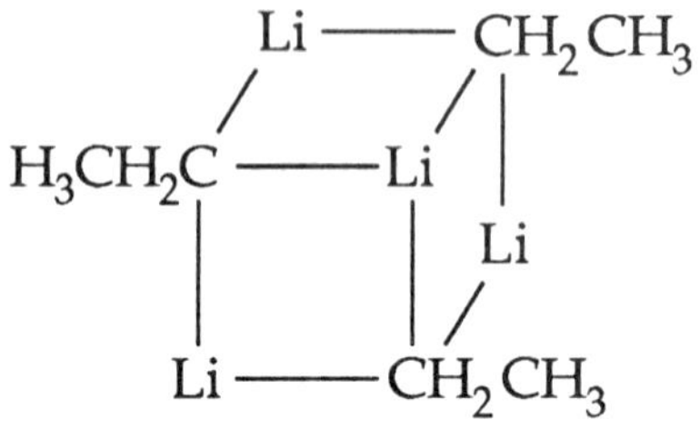

(e) **RbCH3.** This compound is named methylrubidium. It is a saline, electron-deficient methyl compound.

**10.5   The tendency toward association.** For the series of compounds B(CH3)3, Al(CH3)3, Ga(CH3)3, and In(CH3)3, the tendency to form dimers with bridging methyl groups is greatest for aluminum and decreases dramatically down the group. Trimethylborane is a monomer and shows no tendency to dimerize. The difference between trimethylborane and trimethylaluminum may be due to the small size of boron and the relatively short C–B bonds in the former compound. A structure with bridging methyl groups may be prevented from forming in the case of B(CH₃)₃ because of steric hindrance.

**10.6** **Sketch the structures of: (a) Methyllithium.** The structure of this compound, $(LiCH_3)_4$, is based on a cube, with four Li atoms and four C atoms at the eight corners. The structure can also be described as interpenetrating $Li_4$ and $C_4$ tetrahedra, and is shown below.

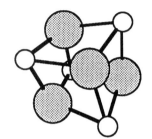

The structure of methyllithium. The larger shaded circles are the Li atoms and the smaller open circles are the C atoms. The H atoms of the methyl groups have been omitted for clarity. The C–Li distances are 2.31 Å (cf. C–B = 1.58 Å in B(CH₃)₃, below).

**(b) Trimethylboron.** This compound contains a trigonal planar B atom with three terminal methyl groups and is shown below.

$$B(CH_3)_3 \qquad Al_2(CH_3)_6 \qquad Si(CH_3)_4$$

**(c) Hexamethyldialuminum.** This compound contains two tetrahedral Al atoms bridged by methyl groups. Each Al atom also has two terminal methyl groups, to give the structure shown above.

**(d) Tetramethylsilane.** This compound contains a tetrahedral Si atom with four terminal methyl groups and is shown above.

**(e) Trimethylarsine.** This compound contains a trigonal pyramidal As atom with three terminal methyl groups, similar to the structure of $AsH_3$. Its structure is shown at the right. The C–As–C bond angles are 96° (the H–As–H bond angles in $AsH_3$ are 91.8° (**Table 9.4**)).

$As(CH_3)_3$

**10.7  Which of the following is a (1) good carbanion reagent, (2) mild Lewis acid, (3) mild Lewis base, and/or (4) strong reducing agent? (a)** $Li_4(CH_3)_4$. Methyl lithium is a good carbanion nucleophile reagent, a mild Lewis acid (the $Li^+$ ions will form complexes with Lewis bases such as tetramethyl-ethylenediamine (see **Section 10.6** and **Structural Drawing 7**), and a strong reducing agent (for example, methyllithium reacts with oxygen to form lithium methoxide, $LiOCH_3$; this is a net reduction of oxygen and oxidation of carbon).

**(b)** $Zn(CH_3)_2$. Dimethylzinc is also a good carbanion reagent, although not as reactive as methyllithium. Recall that zinc alkyls were used quite extensively in organic synthesis until the advent of the more reactive Grignard reagents (see part **(c)**, below). Dimethylzinc is also a mild Lewis acid and a strong reducing agent (recall that it also reacts spontaneously and vigorously with oxygen when it is exposed to air).

**(c)** $(CH_3)MgBr$. Methylmagnesium bromide, like all Grignard reagents, is a good carbanion nucleophile reagent, a mild Lewis acid, and a strong reducing agent.

**(d)** $B(CH_3)_3$. Trimethylborane is a mild Lewis acid, but it is not a good carbanion reagent or a strong reducing agent. While it will react with oxygen like $Li_4(CH_3)_4$, $Zn(CH_3)_2$, and $(CH_3)MgBr$, the central B atom is not as electropositive as Li, Zn, or Mg. Therefore, its organometallic compounds are not as good reducing agents as the organometallic compounds of these three elements.

**(e)** $Al_2(CH_3)_6$. Hexamethyldialuminum (commonly referred to as trimethylaluminum) is a good carbanion nucleophile reagent, a moderately strong Lewis acid, and a strong reducing agent. Aluminum alkyls are in fact used to reduce Ti(IV) compounds to Ti(III) compounds that are olefin polymerization catalysts.

(f) $Si(CH_3)_4$. Tetramethylsilane, like many other electron precise organometallic compounds, exhibits none of the four properties mentioned above. It has no lone pairs, so it does not exhibit Lewis basicity. It does not have a low-lying vacant orbital or cannot dissociate into a fragment that does, so it does not exhibit Lewis acidity. In fact, its most characteristic property is that it is inert, which is one of the main reasons it can be used as a chemical shift standard for $^1H$, $^{13}C$, and $^{29}Si$ NMR spectroscopy (i.e. it can be added to solutions of almost any substance without reaction).

(g) $As(CH_3)_3$. Trimethylarsene (or trimethylarsine) is a mild Lewis base, since the central As atom has a lone pair of electrons. It is not a good carbanion reagent, a strong reducing agent, nor a Lewis acid, since the central As atom is electronegative, not electropositive (compare its electronegativity, 2.18 (Table 1.9), with that of elements whose organometallic compounds are good carbanion reagents, such as Li (0.98), Mg (1.31), Al (1.61), and Si (1.90)).

10.8   Balanced chemical equations. (a) A carbanion reagent with $AsCl_3$ and with $SiPh_2Cl_2$. One reagent we might chose is methyllithium, which in ether solution in the presence of $AsCl_3$ would undergo a metathesis reaction as follows:

$$3LiCH_3 + AsCl_3 \rightarrow 3LiCl \ (s) + As(CH_3)_3$$

Another reagent might be $(CH_3)MgBr$, which in ether solution in the presence of $SiPh_2Cl_2$ would undergo the following metathesis reaction:

$$2(CH_3)MgBr + SiPh_2Cl_2 \rightarrow 2Mg(Br, Cl)_2 + SiPh_2(CH_3)_2$$

Note that these reactions are spontaneous because Li and Mg are more electropositive than As and Si, respectively.

(b) A Lewis acid with $NH_3$. A straightforward reaction would be one between trimethylborane and ammonia, which could be carried out without using a solvent, and which would result in the solid complex shown in the balanced equation below:

$$B(CH_3)_3 \ (g) \ + \ NH_3 \ (g) \ \rightarrow \ (CH_3)_3B-NH_3 \ (s)$$

**(c) A Lewis base with [Hg(CH₃)][BF₄].** The only Lewis base in **End-of-Chapter Excercise 10.7** is trimethylarsine, which would form a complex with the cationic Lewis acid $Hg(CH_3)^+$ as follows:

$$As(CH_3)_3 \ + \ [Hg(CH_3)][BF_4] \ \rightarrow \ [(CH_3)_3As-Hg(CH_3)][BF_4]$$

The structure of the $(CH_3)_3As-Hg(CH_3)^+$ cation, a complex of the Lewis base $As(CH_3)_3$ and the Lewis acid $Hg(CH_3)^+$. Note that the Hg atom is linear two-coordinate.

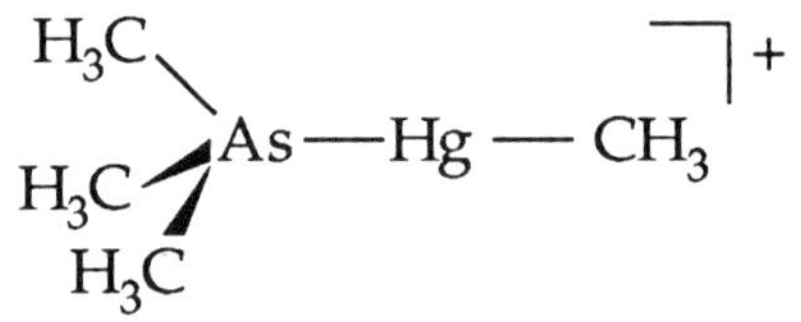

**10.9  Give examples of the following reaction types: (a) A metal with an organic halide.** The formation, stability, and reactivity of organometallic compounds is discussed in **Section 10.3**. One of the fundamental ways to prepare an organometallic compound of an electropositive metal is to react the metal with an alkyl or aryl halide. The net reaction can be either (i) or (ii), below, depending on whether the oxidation state of the metal is normally +1 or higher:

$$2M \ + \ RX \ \rightarrow \ MR \ + \ MX \qquad \qquad \text{(i)}$$

$$M \ + \ RX \ \rightarrow \ RMX \qquad \qquad \text{(ii)}$$

Specific examples are:

$$2Li \ + \ n\text{-BuCl} \ \rightarrow \ n\text{-BuLi} \ + \ LiCl \quad \text{and} \quad Mg \ + \ PhBr \ \rightarrow \ PhMgBr$$

**(b)  Transmetallation.** In this type of reaction, one metal takes the place of another, as in:

$$M + M'R \rightarrow M' + MR$$

The most important factor in determining the course of the reaction is the extent to which M is more electropositive that M'. For example, Al will displace Zn from its organometallic compounds, since Al is more electropositive than Zn, but it will not displace lithium from its organometallic compounds, because Al is less electropositive than Li:

$$2Al\ (s) + 3Zn(C_2H_5)_2 \rightarrow Al_2(C_2H_5)_6 + 3Zn\ (s)$$

$$Al\ (s) + LiC_2H_5 \rightarrow NR$$

(c) **Metathesis.** A metathesis reaction is a double replacement reaction. In the context of organometallic chemistry, it is a reaction involving an organometallic MR or $MR_n$ compound and a halide of some element, such as $EX_n$ or $R'_nEX$. So long as M is more electropositive than E, the following reaction will take place:

$$M–R + E–X \rightarrow M–X + E–R$$

Specific examples are:

$$3LiC_3H_7 + PCl_3 \rightarrow 3LiCl + P(C_3H_7)_3$$

$$LiC_3H_7 + Si(CH_3)_3Cl \rightarrow LiCl + Si(CH_3)_3(C_3H_7)$$

$$3PhMgCl + BCl_3 \rightarrow 3MgCl_2 + BPh_3$$

$$PhMgCl + BCl_3 \rightarrow MgCl_2 + BPhCl_2$$

The first example illustrates the way that most alkyl and aryl phosphines are prepared (these are very important ligands for transition metals, as you will see later). The second example demonstrates the utility of metathesis reactions for the preparation of *p*-block organometallics containing more than one type of alkyl group. The third and fourth examples show that, in many cases, controlling the stoichiometry of the reaction can affect the composition of the final product.

**10.10 Which compound is likely to be the stronger reducing agent?** (a) **Na[C$_{10}$H$_8$] or Na[C$_{14}$H$_{10}$].** Sodium naphthalide and sodium anthracenide are examples of organometallic salts with a delocalized anion. If you remember the structures of naphalene and anthracene, shown below, you can appre-

naphthalene  anthracene

appreciate the fact that naphthalene has a smaller $\pi$ system and hence a more negative reduction potential than anthracene (see **Table 10.3**; remember that the electron given up by sodium and taken up by the organic molecule is added to the lowest unoccupied $\pi^*$ orbital). For this reason, the anion radical of naththalene will give up its extra electron more readily than the anion radical of anthracene. Thus, sodium naphthalide is the stronger reducing agent.

(b) **Na[C$_{10}$H$_8$] or Na$_2$[C$_{10}$H$_8$].** In this case the organic molecule is the same, but the charge is different. That is, in Na[C$_{10}$H$_8$] the organic species is a radical anion while in Na$_2$[C$_{10}$H$_8$] it is a dianion. In almost every case, the first reduction potential of a neutral molecule will be less negative than the second reduction potential (the few exceptions are species that undergo a single two–electron reduction, a rare occurance). For this reason, the dianion will give up an electron more readily than the radical anion. (Compare this situation to the first and second electron affinities of an O atom, listed in **Table 1.8**.) Thus, Na$_2$[C$_{10}$H$_8$] is the stronger reducing agent. As far as the synthesis of Na$_2$[C$_{10}$H$_8$] is concerned, the relative reduction potentials require that excess sodium, and not excess naphthalene, be present. Otherwise, the conproportionation reaction below would occur:

$$Na_2[C_{10}H_8] + C_{10}H_8 \rightarrow 2Na[C_{10}H_8]$$

**10.11 Balanced chemical equations and reaction type.** (a) **Ca with Hg(CH$_3$)$_3$.** Whenever a metal, in this case Ca, is mixed with an organometallic compound, in this case dimethylmercury, a transmetallation can potentially take place. Remember that a more electropositive element will replace a less electropositive element, and since Ca is more electropositive than Hg, the following reaction will take place.

$$Ca\ (s)\ +\ Hg(CH_3)_2\ \rightarrow\ Ca(CH_3)_2\ +\ Hg\ (l)$$

**(b) Hg with $Zn(C_2H_5)_2$.** Once again a transmetallation is the potential reaction, since a metal and an organometallic compound are mixed together. However, since Hg is less electropositive than Zn, no reaction will take place:

$$Hg\ (l)\ +\ Zn(C_2H_5)_2\ \rightarrow\ NR$$

**(c) $LiCH_3$ with $SiPh_3Cl$ in ether.** Whenever an organometallic compound and a halide of some element are mixed together, a metathesis reaction is the possible occurance (see the answer to **End-of-Chapter Exercise 10.9(c)**, above). The metathesis will occur if the organometallic metal atom (Li in this case) is more electropositive than the element of the halide compound (Si in this case). Since Li is far more electropositive than Si, the metathesis reaction will occur, as shown below:

$$LiCH_3\ +\ SiPh_3Cl\ \rightarrow\ LiCl\ +\ SiPh_3(CH_3)$$

**(d) $Si(CH_3)_4$ with $ZnCl_2$ in ether.** As in part **(c)**, above, we have mixed an organometallic compound with a halide. However, in this case the organometallic compound contains Si, which is less electropositive than the element in the halide, namely Zn. Therefore, no reaction will occur:

$$Si(CH_3)_4\ +\ ZnCl_2\ \rightarrow\ NR$$

**(e) $SiH(CH_3)_3$ with $C_2H_4$.** In this case, none of the reaction types of organometallic compounds we have studied is evident. Since a metal is not present, we cannot have a transmetallation. Since a halide is not present, we cannot have a metathesis reaction. However, before you conclude that no reaction will occur, consider that the presence of C–M bonds in a compound might not change the reactivity of other portions of the molecule. In other words, even though you cannot think of a reaction that will cleave the C–Si bonds, you must consider reactions that might cleave the H–Si bond. Recall that reagents containing Si–H bonds can add across double bonds in reactions called hydrosilations (see the end of **Section 9.13**). The platinic acid is added to the reaction medium as a catalyst, so the net reaction involves only the silane and the olefin:

$$SiH(CH_3)_3 + C_2H_4 \rightarrow Si(C_2H_5)(CH_3)_3$$

**10.12 Give balanced chemical equations for the following: (a) Direct synthesis of $Si(CH_3)_2Cl_2$.** The reaction of two equivalents of $CH_3Cl$ with elemental silicon can be carried out at high temperatures in the presence of a catalyst (Cu is generally used):

$$Si\ (s)\ +\ 2CH_3Cl\ (g)\ \rightarrow\ Si(CH_3)_2Cl_2\ (g)$$

**(b) Redistribution of $Si(CH_3)_2Cl_2$.** The exact ratio of products are difficult to predict, because subtle aspects of bonding and steric interactions affect the relative thermodynamic stabilities of the various species $Si(CH_3)_xCl_{4-x}$ ($x = 0 - 4$). A nonbalanced equation is:

$$Si(CH_3)_2Cl_2 \rightarrow Si(CH_3)_4 + Si(CH_3)_3Cl + Si(CH_3)Cl_3 + SiCl_4$$

**10.13 The synthesis of poly(dimethylsiloxane).** Two different kinds of siloxanes are generally used, a cyclic compound like $[(CH_3)_2SiO]_4$, with two methyl groups and two oxygen atoms per silicon atom, and a noncyclic compound such as $((CH_3)_3Si)_2O$, with three methyl groups and only one oxygen atom per silicon atom. These compounds are prepared as follows:

$$Si\ (s)\ +\ 2CH_3Cl\ \rightarrow\ Si(CH_3)_2Cl_2 \quad \text{(formation of C–Si bonds)}$$

$$4Si(CH_3)_2Cl_2 + 4H_2O \rightarrow [(CH_3)_2SiO]_4 + 8HCl \quad \text{(hydrolysis rxn)}$$

$$2Si(CH_3)_2Cl_2 \rightarrow Si(CH_3)_3Cl + Si(CH_3)Cl_3 \quad \text{(redistribution rxn)}$$

$$2Si(CH_3)_3Cl + H_2O \rightarrow ((CH_3)_3Si)_2O + 2HCl \quad \text{(hydrolysis rxn)}$$

Then a mixture of $[(CH_3)_2SiO]_4$ and $((CH_3)_3Si)_2O$ is polymerized using a sulfuric acid catalyst:

$$n[(CH_3)_2SiO]_4 + ((CH_3)_3Si)_2O \rightarrow (CH_3)_3SiO[Si(CH_3)_2O]_{4n}Si(CH_3)_3$$

**10.14 Trends in the oxidation states for Group 13 and Group 15 Organometallics.** For the elements of Groups 13 and 14, the general trends exhibited by all of their compounds is also observed for their organometallic derivatives. That is, the maximum oxidation states of +3 and +4, respectively, are the only important oxidation states at the top of the group, and oxidation states +1 and +2, respectively, are only important at the bottom of each group. In other words, the only organometallic derivatives of boron, aluminum, and silicon are $BR_3$, $AlR_3$, and $SiR_4$; no compounds of composition $BR$, $AlR$, or $SiR_2$ are known. At the other extreme, stable organometallic derivatives of thallium have formulas $TlR$ and $TlR_3$, and tin and lead form both divalent and tetravalent organometallic compounds having formulas $SnR_2$ and $PbR_2$, and $SnR_4$ and $PbR_4$, respectively. The elements of Group 15, As, Sb, and Bi, exhibit a parallel but slightly different trend. Oxidation state +5 is found in some organometallic compounds of As and Sb, such as $[AsPh_4]Br$ and $SbPh_5$, but this oxidation state is not as important as +3, even for arsenic.

**10.15 Describe the periodic trends for the following: (a) Metal-carbon bond enthalpies.** These exhibit a uniform decrease going down a group (see **Table 10.2**), similar to the trend exhibited by the element-hydrogen bond energies for *p*-block elements, which was discussed in **Section 9.8**. Recall that the weak bonds formed by the heavier element near the bottom of each *p*-block group is attributed to poor overlap with the diffuse *s* and *p* orbitals of these very large atoms.

**(b) Lewis acidity.** With one very important exception, Lewis acidity decreases down a group. As discussed in part **(a)**, above, bond enthalpies decrease down each group, including the enthalpy of the bond between the Lewis acid and the base. The exception is the Period 2 element in each group, which is so small that steric hindrance can prevent a strong complex from forming between a Period 2 organometallic Lewis acid and a base. For example, aluminum alkyls are the strongest Lewis acids among Group 13 organometallics. Boron alkyls are considerably weaker acids (see below).

Lewis acidity:    $BR_3 < AlR_3 > GaR_3 > InR_3 > TlR_3$

(c)   **Hydrolytic stability.**   The organometallic compounds of eletropositive elements, such as those of Groups 1, 2, 12, and 13, react with water.   However, within each group, the tendency to hydrolyze decreases down the group.   This is because the heavier members of each group are less electropositive than the lighter members, and the carbanionic character of the organo groups decreases as the metal becomes less electropositive.   Thus, while $ZnR_2$ and $AlR_3$ compounds completely hydrolyze in water, compounds containing the $HgR^+$ and $GaR_2^+$ groups are stable in water.   The organometallic compounds of the Group 14 and 15 elements do not hydrolyze in water.

$$H_3C-Tl-CH_3 \quad \bigg]^+$$

The $Tl(CH_3)_2^+$ cation, which is stable in aqueous solution and is linear like the isoelectronic species $Hg(CH_3)_2$.

**10.16   Summarize the trend in each of the following: (a)   The relative ease of pyrolysis of $Si(CH_3)_4$ and $Sn(CH_3)_4$.**   Due to the relative bond strengths C–Sn < C–Si, tetramethyltin will undergo thermal decomposition (pyrolysis) more rapidly at 300 °C than will tetramethylsilane.

Rate of thermal decomposition:     $Sn(CH_3)_4 > Si(CH_3)_4$

(b)   **The relative Lewis acidity of $Li_4(CH_3)_4$, $B(CH_3)_3$, $Si(CH_3)_4$, and $Si(CH_3)Cl_3$.**   Tetramethylsilane exhibits no Lewis acidity, so it is by default the weakest Lewis acid of these four compounds.   The presence of the three chlorine ligands in $Si(CH_3)Cl_3$ renders the silicon center more acidic than $Si(CH_3)_4$, but it is still a relatively weak Lewis acid.   Considering the other two, trimethylboron is commonly referred to as a weak acid, but it is stronger than methyllithium.   Recall that the synthesis of the tetraphenylborate anion may be viewed as the transfer of the strong base $Ph^-$ from the weak Lewis acid $Li^+$ to the stronger acid $BPh_3$:

$$BPh_3 + LiPh \rightarrow Li[BPh_4]$$

Therefore, the order of Lewis acidity is $Si(CH_3)_4 < Si(CH_3)Cl_3 < Li_4(CH_3)_4 < B(CH_3)_3$.

(c) **The relative Lewis basicity of Si(CH₃)₄ and As(CH₃)₃.** For a substance to be basic, it must have either one or more lone pairs of electrons or one or more loosely held bonding pairs of electrons (generally these are π electrons). Tetramethylsilane has neither of these, so it is not basic at all. On the other hand, trimethylarsine has a lone pair of electrons on the central As atom, rendering it a mild Lewis base.

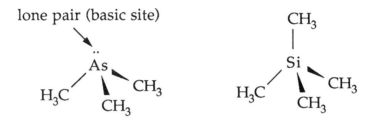

(d) **The tendency of Li₄(CH₃)₄ and Hg(CH₃)₂ to displace halide from GeCl₄.** A halide displacement by an organometallic compound is an example of a metathesis reaction. You should recall that in this type of metathesis reaction, the more electropositive element will wind up with the halide, and the less electropositive element will wind up with the organic group. Methyllithium *will* displace chloride from tetrachlorogermane, because lithium is more electropositive than germanium, and dimethylmercury *will not* displace chloride from tetrachlorogermane, because mercury is less electropositive than germanium:

$$Li_4(CH_3)_4 + GeCl_4 \rightarrow 4LiCl + Ge(CH_3)_4$$

$$Hg(CH_3)_2 + GeCl_4 \rightarrow NR$$

**10.17 What air-free technique is appropriate for handling: (a) Li₄(CH₃)₄ in ether?** Methyllithium is sensitive to both oxygen and water, so it cannot be handled in open flasks. Generally, Schlenk apparatus, air-tight syringes, and stainless steel cannula are used to handle air-sensitive solutions (see **Box 10.1**). Alternatively, solutions can be handled in an inert-atmosphere glovebox, but this is usually not as convenient since a glovebox is a relatively elaborate piece of equipment and requires constant maintenance. Gloveboxes are ideal for handling nonvolatile air-sensitive solids.

**(b) Trimethylboron?** Since this substance is a pyrophoric gas at room temperature, it is most conveniently handled using a vacuum line (pyrophoric: capable of igniting spontaneously when exposed to air). Of all the air-free techniques mentioned in the text, a vacuum line offers the best protection from the atmosphere. In general, however, it cannot be used to handle solutions, since the solvent and solute will usually have different volatilities (in some cases the solute will be nonvolatile).

$$B(CH_3)_3 \quad mp -162\ °C, bp -20\ °C$$

**(c) Triisobutylaluminum?** This substance is a pyrophoric liquid and is not volatile enough to transfer from one bulb to another using a vacuum line. Schlenk apparatus, air-tight syringes, and stainless steel cannula are used to transfer $Al(i-Bu)_3$ and other aluminum alkyls from one container to another. Similar inert-atmosphere techniques are used to handle industrial sized amounts of these compounds: the "flasks" can be as large as railroad tank cars.

$$Al_2(i-Bu)_6 \quad mp\ 4\ °C, bp\ 73\ °C\ at\ 5\ Torr$$

**(d) Solid AsPh₃?** Triphenylarsine is not oxygen- or water-sensitive, so it can be handled in an open flask. If it is to be added to an air-sensitive reaction mixture, it can be weighed out in the air and transfered to a Schlenk flask, which is then purged with an inert gas. At this point, an air-sensitive liquid compound or solution can be added using cannula.

$$AsPh_3 \quad mp\ 61\ °C, bp\ 233\ °C\ at\ 14\ Torr$$

**(e) Liquid $(CH_3)_3SiOSi(CH_3)_3$?** Hexamethyldisiloxane is a liquid that is not oxygen- or water-sensitive. It can be handled in an open flask. Since it is relatively volatile (its boiling point is 101 °C), it can be added to an air-sensitive reaction mixture by any of the air-free techniques, including a vacuum line. To help develop a "feel" for relative volatilities, consider the following:

$$Si(CH_3)_4 \quad \text{b.p. } 27\,°C \qquad\qquad CH_4 \quad \text{b.p. } -164\,°C$$

$$(CH_3)_3SiOSi(CH_3)_3 \text{ b.p. } 101\,°C \qquad H_3COCH_3 \quad \text{b.p. } -23\,°C$$

**10.18 The electronic structure of $Si_6(CH_3)_{12}$.** As discussed in **Section 10.12**, the near-UV absorption band is indicative of a relatively small HOMO-LUMO gap, which is a manifestation of a delocalized low-lying vacant orbital. Recall that *unsaturated* hydrocarbons such as naphthalene, which also have a small HOMO-LUMO gap, are reduced by sodium to form anion radicals (see **End-of-Chapter Exercise 10.10**). You should recognize that this behavior of cyclic silanes is in sharp contrast with their carbon analogues, cyclohexane and substituted cyclohexanes, which do not have near-UV absorptions. The ability of $Si_6(CH_3)_{12}$ to form a stable anion such as $[Si_6(CH_3)_{12}]^-$ upon treatment with sodium metal is also in sharp contrast with the behavior of cyclohexanes, which do not react with sodium.

**10.19 The synthesis of $R_2Si=SiR_2$.** As discussed in **Section 10.12**, disilenes can be formed by the photochemical cleavage of two of the Si–Si single bonds in a cyclic trisilane:

It will be of paramount importance to use R groups that are very bulky so that the disilene does not dimerize or polymerize. So, now we must develop a synthesis for a cyclic trisilane with bulky substituents starting with $SiR_2Cl_2$. A convenient and general route for the formation of element–element single bonds of the *p*-block elements is reduction of a halide of that element with an alkali metal or with another strong reducing agent. For example, $(CH_3)_3SiSi(CH_3)_3$ can be formed by reacting $Si(CH_3)_3Cl$ with sodium as follows:

$$2Si(CH_3)_3Cl + 2Na\,(s) \rightarrow (CH_3)_3SiSi(CH_3)_3 + 2NaCl\,(s)$$

This provides the necessary analogy for a route to our cyclic compound:

$$3SiR_2Cl_2 + 6Na \ (s) \rightarrow cyclo\text{-}Si_3R_6 + 6NaCl \ (s)$$

# CHAPTER 11

# THE BORON AND CARBON GROUPS

| 12 | 13 | 14 | 15 |
|----|----|----|----|
|    | B  | C  | N  |
|    | Al | Si | P  |
| Zn | Ga | Ge | As |
| Cd | In | Sn | Sb |
| Hg | Tl | Pb | Bi |
|    | III | IV |   |

The elements of the boron and carbon groups exhibit a wide range of chemical and physical properties. Examples of nonmetals, semimetals, and metals can be found in the two groups. Many of the elements have two or more polymorphs, including B (several allotropes), C (graphite and diamond), and Sn (grey and white). The electronic properties of some of the elements and their compounds are of current technological importance.

## SOLUTIONS TO IN-CHAPTER EXERCISES

**11.1 Balanced equations for the following reaction mixtures: (a) $BCl_3$ and ethanol.** As mentioned in the **Example**, boron trichloride is vigorously hydrolyzed by water. Therefore, a good assumption is that it will also react with protic solvents such as alcohols, forming HCl and B–O bonds:

$$BCl_3\ (g)\ +\ 3EtOH\ (l)\ \rightarrow\ B(OEt)_3\ (l)\ +\ 3HCl\ (g)$$

**(b) BCl₃ and pyridine in hydrocarbon solution.** Neither pyridine nor hydrocarbons can cause the protolysis of the B–Cl bonds of boron trichloride, so the only reaction that will occur is a complex formation reaction:

$$BCl_3 \ (g) + py \ (l) \rightarrow Cl_3B{-}py \ (s)$$

Note that only a 1:1 complex is formed, even if excess pyridine (py) is used. Boron and the other Period 2 atoms cannot become hypervalent, in constrast with the heavier atoms of Periods 3 and beyond.

**(c) BBr₃ and F₃BN(CH₃)₃.** Since boron tribromide is a stronger Lewis acid than boron trifluoride, it will displace $BF_3$ from its complex with $N(CH_3)_3$:

$$BBr_3 \ (l) + F_3BN(CH_3)_3 \ (s) \rightarrow BF_3 \ (g) + Br_3BN(CH_3)_3 \ (s)$$

**11.2   Synthesis of (CH₃)₃B₃N₃(CH₃)₃.** This compound is permethyl-borazine. As discussed in **Section 11.3**, the reaction of ammonium chloride with boron trichloride yields *B*-trichloroborazine, while the reaction of a primary ammonium chloride with boron trichloride yields *N*-alkyl substituted B-trichloroborazine, as shown below:

$$3RNH_3{}^+Cl^- + 3BCl_3 \rightarrow 9HCl + Cl_3BNR_3 \ (R = H, \ alkyl)$$

Therefore, if we use methylammonium chloride we will produce $Cl_3BN(CH_3)_3$ (i.e. R = $CH_3$). This product can be converted to the desired one by treating it with an organometallic methyl compound of a metal that is more electropositive than boron (you should review metathesis reactions involving organometallic compounds, discussed in **Section 10.3**). Either methyllithium or methylmagnesium bromide could be used:

$$Cl_3BN(CH_3)_3 + 3CH_3MgBr \rightarrow (CH_3)_3B_3N_3(CH_3)_3 + 3Mg(Br, \ Cl)_2$$

The structure of $N, N', N''$-trimethyl-
$B, B', B''$-trimethylborazine.

**11.3  Propose chemical reactions between: (a) $(CH_3)_3NBF_3$ and $GaCl_3$.**
Towards a hard base such as trimethylamine, the order of Lewis acidity is
$BCl_3 > GaCl_3$, so it is a good guess that the order $BF_3 > GaCl_3$ would be found
as well.  On this basis, we can rule out a simple displacement (i.e. the
products $BF_3$ and $(CH_3)_3NGaCl_3$ will *not* be formed).  However, given that
the lattice energy of $GaF_3$ is large (this is mentioned in the **Example**), a good
proposal is that the following halide exchange reaction will occur:

$$(CH_3)_3NBF_3 + GaCl_3 \rightarrow (CH_3)_3NBCl_3 + GaF_3$$

Not only will the reaction be driven by the greater lattice energy of $GaF_3$ as
compared with $GaCl_3$, but the N–B bond in the Lewis acid-base complex will
be stronger for the $BCl_3$ complex than for the $BF_3$ complex, since $BCl_3$ is the
stronger Lewis acid.

**(b) $TlCl_3$ and formaldehyde in acidic water.**  Towards the bottom of
Groups 13/III and 14/IV, oxidation states two lower than the maximum
oxidation state become increasingly more stable and hence more important.
This being the case, it should come as no surprise than $TlCl_3$ is highly
oxidizing (i.e. it is readily reduced to $TlCl$).  Since formaldehyde is easily
oxidized, the following redox reaction will occur:

$$2TlCl_3\ (aq) + CH_2O\ (aq) + H_2O\ (l) \rightarrow 2TlCl\ (s) + CO_2\ (aq) + 4HCl\ (aq)$$

This is the net reaction of the following two half–reactions:

$$2TlCl_3\ (aq) + 4e^- \rightarrow 2TlCl\ (s) + 2Cl^-\ (aq)$$

$$CH_2O\ (aq) + H_2O\ (l) \rightarrow CO_2\ (aq) + 4H^+\ (aq) + 4e^-$$

Note that TlCl is insoluble in water, like AgCl. In many reactions, $Tl^+$ ion can be used to precipitate chloride, bromide, or iodide, just like $Ag^+$.

**11.4 Describe how the electronic structure of graphite is altered when it reacts with: (a) Potassium.** The extended $\pi$ system for each of the planes of graphite result in a band of $\pi$ orbitals. The band is half filled in pure graphite — that is, all of the bonding MOs are filled and all of the antibonding MOs are empty. The HOMO-LUMO gap is ~0 eV, giving rise to the observed electrical conductivity of graphite. Chemical reductants, like potassium, can donate their electrons to the LUMOs (graphite $\pi^*$ orbitals), resulting in a material with a higher conductivity.

    **(b) Bromine.** Chemical oxidants, like bromine, can remove electrons from the $\pi$-symmetry HOMOs of graphite. This also results in a material with a higher conductivity.

**11.5 Determine the charge on $[Si_4O_{12}]^{n-}$.** Two views of the structure of this cyclic silicate are shown below. It is an eight-membered ring of alternating Si

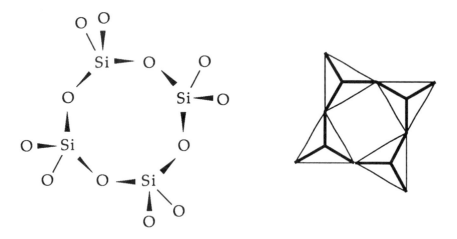

and O atoms with eight terminal Si–O bonds, two on each Si atom. Since each terminal O atom contributes –1 to the total charge on the cyclic ion, the overall charge is –8. The charge can also be determined from the oxidation numbers of the elements: $4(+4) + 12(-2) = -8$.

**11.6    How many Si and Al atoms are in one sodalite cage?** A sodalite cage is based on a truncated octahedron. We can imagine an octahedron with four of its six vertices in the plane of this page. We would then look down upon one of the remaining vertices. This is the shown in the drawing below and on the left. The truncation we shall make is parallel to the plane of the page.

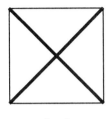

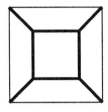

octahedron                    octahedron truncated
                              along one of its $C_4$ axes

The top vertex is removed, and in its place is a square plane parallel to the plane of the page. This is shown in the drawing on the right. Four new vertices have been created for the one that was removed. If we imagine performing this procedure six times, once for each of the vertices of the original octahedron, we will remove six vertices and create 24 new ones (6 x 4). Thus, a sodalite cage has 24 Si and Al atoms.

**11.7    How many skeletal electrons are present in $B_5H_{11}$?** Five B–H units contribute 5 x 2 = 10 electrons and the six additional H atoms contribute six additional electrons, for a total of 16 electrons or 8 pairs of electrons.

**11.8    Use Wade's rules to determine the structure of $B_5H_9$.** Five B–H units contribute 5 x 2 = 10 electrons and the four additional H atoms contribute four additional electrons, for a total of 14 electrons or 7 pairs of electrons. Boranes of formula $B_nH_{n+4}$ have the *nido* structure, which is based on a *closo* structure with n + 1 vertices. In this case n = 5. The *closo* structure with 6 vertices is an octahedron, so the *nido* structure of $B_5H_9$ is based on an octahedron with one vertex missing, which is a square pyramid. In this compound, the four additional H atoms bridge the four B atoms that comprise the square plane of the square pyramid. Wade's rules do not predict the positions of the additional H atoms, just the framework structure of the compound. The $C_{4v}$ structure of $B_5H_9$ is shown in **Figure 11.7**.

**11.9**   **Propose a synthesis for 1,7-$B_{10}H_{10}C_2(Si(CH_3)_2Cl)_2$.** As in the **Example**, you should consider attaching the $-Si(CH_3)_2Cl$ substituents to the carbon atoms of this carborane by using the dilithium derivative 1,7-$B_{10}H_{10}C_2Li_2$. You can first prepare 1,2-$B_{10}C_2H_{12}$ from decaborane as in the **Example**. Then, this compound is thermally converted to a mixture of the the 1,7- and 1,12-isomers, which can be separated by chromatography:

$$1,2\text{-}B_{10}C_2H_{12} \xrightarrow{\sim 500\ °C} 1,7\text{-}B_{10}C_2H_{12}\ (90\%)\ +\ 1,12\text{-}B_{10}C_2H_{12}\ (10\%)$$

The pure 1,7-isomer is lithiated with RLi and then treated with $Si(CH_3)_2Cl_2$:

$$1,7\text{-}B_{10}H_{10}C_2Li\ +\ 2Si(CH_3)_2Cl_2\ \rightarrow\ 1,7\text{-}B_{10}H_{10}C_2(Si(CH_3)_2Cl)_2\ +\ 2LiCl$$

---

## SOLUTIONS TO END-OF-CHAPTER EXERCISES

**11.1**   **General properties of the elements.**

|   | type of element | diamond structure? | primarily occurs as oxide(s)? | also occurs as sulfide(s)? |
|---|---|---|---|---|
| B  | nonmetal | no  | yes | no  |
| Al | metal    | no  | yes | no  |
| Ga | metal    | no  | yes | yes |
| In | metal    | no  | no  | yes |
| Tl | metal    | no  | no  | yes |
|    |          |     |     |     |
| C  | nonmetal | yes | yes | no  |
| Si | nonmetal | yes | yes | no  |
| Ge | nonmetal | yes | yes | no  |
| Sn | metal    | yes | yes | yes |
| Pb | metal    | no  | no  | yes |

**11.2   Describe the repeating unit in the structures of B and Ga.** In the different polymorphic forms of elemental boron, $B_{12}$ icosahedra are linked together in various ways. Gallium has an unusual structure. Despite its metallic character, it does not have a close-packed structure like its congeners indium and thallium, and only has 7 nearest neighbors instead of 12. The unusual structure of gallium is certainly a factor in its unusually low melting point (30 °C).

**11.3   The synthesis of: (a) Aluminum chloride.** The apparatus that is normally used for the synthesis of anhydrous $AlCl_3$ is shown in **Figure 11.5**. It is called a hot tube reactor. Flowing HCl gas is heated and passed over a sample of pure Al metal. The product, $AlCl_3$, is somewhat volatile and sublimes out of the hot tube, condensing on the cold walls of the glass tube. The balanced equation is:

$$2Al \ (s) \ + \ 6HCl \ (g) \ \rightarrow \ 2AlCl_3 \ (s) \ + \ 3H_2 \ (g)$$

**(b)   Tin(II) chloride.** The same apparatus can be used with Sn metal in place of Al metal. The balanced equation is:

$$Sn \ (s) \ + \ 2HCl \ (g) \ \rightarrow \ SnCl_2 \ (s) \ + \ H_2 \ (g)$$

**11.4   Lewis acidity of Group 14 halides. (a) Arrange in order of increasing Lewis acidity toward hard Lewis bases: $BF_3$, $BCl_3$, $SiF_4$, $AlCl_3$.** For a given halogen, the order of acidity for Group 13 halides toward hard Lewis bases like dimethylether or trimethylamine is $BX_3 > AlX_3 > GaX_3$, while the order toward soft Lewis bases such as dimethylsulfide or trimethylphosphine is $BX_3 < AlX_3 < GaX_3$. This fact establishes the order $BCl_3 > AlCl_3$ for the four Lewis acids in question. For boron halides, the order of acidity is $BF_3 < BCl_3 < BBr_3$, exactly opposite to the order expected from electronegativity trends. This is discussed in **Section 11.3**. While you can now predict with some confidence that both $BF_3$ and $AlCl_3$ are both weaker Lewis acids than $BCl_3$ towards hard Lewis bases, it is not possible to predict from the information given in the text whether $BF_3$ is stronger or weaker than $AlCl_3$. Finally, silicon tetrahalides are only mild Lewis acids, so $SiF_4$ is the weakest of the four acids given. So, the order of increasing Lewis acidity towards hard Lewis bases is $SiF_4 < BF_3 \sim AlCl_3 < BCl_3$.

**(b) Predict the course of the following reactions:**

(i)   $F_4Si–N(CH_3)_3 + BF_3 \rightarrow F_3B–N(CH_3)_3 + SiF_4$     $(BF_3 > SiF_4)$

(ii)   $F_3B–N(CH_3)_3 + BCl_3 \rightarrow Cl_3B–N(CH_3)_3 + BF_3$     $(BCl_3 > BF_3)$

(iii)   $Cl_3Al–N(CH_3)_3 + BCl_3 \rightarrow Cl_3B–N(CH_3)_3 + AlCl_3$     $(BCl_3 > AlCl_3)$

**11.5   The elements that form saline, metallic, and metalloid carbides.** The distribution of carbides in the periodic table is shown in **Figure 11.20**. If you were not able to completely construct this diagram, try to remember that the saline carbides are formed by the same elements that form saline hydrides and ionic organometallic compounds, namely the alkali metals and the alkaline earths. In addition, aluminum carbide is saline. Metallic carbides are formed by the early and middle $d$-block metals. The metalloid carbides are formed by those two nonmetallic elements that also form organometallic compounds, namely boron and silicon.

**11.6   Describe the preparation, structure, and classification of: (a) KC$_8$.** This compound is formed by heating graphite with potassium vapor or by treating graphite with a solution of potassium in liquid ammonia. The potassium atoms are oxidized to $K^+$ ions; their electrons are added to the LUMO $\pi^*$ orbitals of graphite. The $K^+$ ions intercalate between the planes of the reduced graphite, so that there is a layered structure of alternating $sp^2$ carbon atoms and potassium ions. The structure of $KC_8$ which is an example of a saline carbide, is shown in **Figure 11.21**.

**(b) CaC$_2$.** There are two ways of preparing calcium carbide, and both require very high temperatures ($\geq 2000\ °C$). The first is the direct reaction of the elements, while the second is the reaction of calcium oxide with carbon:

$$Ca\ (l) + 2C\ (s) \rightarrow CaC_2\ (s)$$

$$CaO \ (s) \ + \ 3C \ (s) \ \rightarrow \ CaC_2 \ (s) \ + \ CO \ (g)$$

The structure is quite different from that of $KC_8$. Instead of every carbon atom bonded to three other carbon atoms, as in graphite and $KC_8$, calcium carbide contains discrete $C_2^{2-}$ ions with carbon-carbon triple bonds.

**11.7 Describe the structures of SnO and red PbO.** Tin(II) oxide is isostructural with red lead(II) oxide, the structure of which is shown in **Figure 11.10**. Each metal atom has four oxide nearest neighbors, but these are not symmetrically distributed around the metal atoms. Instead the metal and four oxygen atoms form a square pyramid, with the metal atom at the apex. A structure such as this, with an apparent gaping hole in the metal atom's coordination sphere, usually signals a stereochemically active lone pair of electrons. The electron configurations of $Sn^{2+}$ and $Pb^{2+}$ are $[Kr]4d^{10}5s^2$ and $[Xe]5d^{10}6s^2$, respectively, and it is believed that the $s^2$ pair of electrons is not spherically distributed around the metal but is directed away from the four oxide ion ligands (i.e. along the $C_4$ axis of the square pyramid).

**11.8 List four examples of amorphous or partially crystalline solids.** Three forms of carbon have a low degree of crystallinity. These are carbon black, activated carbon, and carbon fibers, and are discussed in **Section 11.6**. Carbon black is used as a pigment and as an additive to automobile tires to improve their strength and wear resistance. Activated carbon is used as an adsorbent. Carbon fibers have great tensile strength, and so they increase the strength of materials to which they are added. A material referred to as amorphous silicon is really a solid silicon hydride $SiH_x$ ($x \le 0.5$), and was mentioned in **Section 11.7** and was discussed in **Section 9.13**. It is used in photovoltaic devices. A fifth amorphous material mentioned in this chapter is fused quartz (amorphous $SiO_2$, see **Section 11.9**), which is used to manufacture laboratory glassware that is transparent to UV radiation.

**11.9 Differences between B and Al and between C and Si. (a) Structures and electrical properties.** The various allotropes of elemental boron consist of icosahedral $B_{12}$ cages connected together in various ways. In contrast, aluminum is a cubic close-packed metal at all temperatures. In harmony with the more localized bonding in boron, it is a semiconductor whereas aluminum is a metal. The only important form of silicon has the diamond structure, named after one of the naturally occuring forms of carbon. Despite the similar structures, silicon is a semiconductor and the diamond allotrope of carbon is an insulator (however, recall that a semiconductor and an insulator are merely extremes of the same type of conductivity behavior: see **Section 3.4**). The slightly more stable form of carbon, graphite, has no silicon homologue. Graphite is a semiconductor in the direction perpendicular to the planes of $sp^2$ carbon atoms and a metal in the directions parallel to the planes.

**(b) Physical properties and structures of $CO_2$ and $SiO_2$.** Carbon dioxide is a gas at ambient conditions while silicon dioxide is a hard solid with a high melting point (1710 °C). Both compounds contain four covalent bonds to oxygen atoms, but the structures are quite different, as expected from the different physical properties. Whereas $CO_2$ possesses two C=O double bonds, resulting in a triatomic molecule, $SiO_2$ possesses four Si–O single bonds, resulting in an extended three-dimensional covalent lattice. The reasons behind this difference was discussed in **Section 2.8**: Two C–O single bonds are weaker than one C=O double bond, while Two Si–O single bonds are stronger than one Si=O double bond.

The different structures of $CO_2$ and $SiO_2$

**(c)   The Lewis acid/base properties of $CX_4$ and $SiX_4$.** Carbon tetrahalides are not Lewis acids, despite the fact the the central carbon atom has four electronegative ligands. With its valence $s$ and $p$ orbitals taken up by bonding with halogen atoms, carbon has no low lying empty orbitals with which it can accept an electron pair from a Lewis base. In contrast, silicon has valence $3d$ orbitals and can form complexes with one or two Lewis bases, yielding structures in which silicon is 5- or 6-coordinate, respectively.

**(d) The structures of $BX_3$ and $AlX_3$.** The boron halides are trigonal planar molecules that are gases or volatile liquids at ambient conditions. In contrast, aluminum halides are solids with extended lattices for X = F and Cl and with dimeric $Al_2X_6$ molecules for X = Br and I. The differences between the halides of boron and aluminum can be traced to their relative sizes. Since boron is so small it (i) resists formation of the halide bridges present in the $AlX_3$ structures and (ii) can form bonds with halogens that have partial double bond character, in marked contrast with Al–X bonds, which have little or no double bond character (as in the discussion of $CO_2$ vs. $SiO_2$, above, double bonds to Period 3 atoms are very weak).

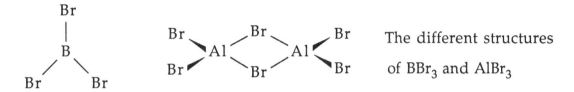

The different structures of $BBr_3$ and $AlBr_3$

**11.10 Stable oxidation states and the inert pair effect.** The lightest members of these groups can exhibit a number of stable oxidation states (e.g., $BCl_3$, $B_2Cl_4$, and $B_4Cl_4$; $CO_2$ and CO). However, the only oxidation states observed for Al and Si are Al(III) and Si(IV). In contrast, the heavier members of each group exhibit a stable oxidation state that is two lower than the maximum oxidation state. The lower down the group, the more stable the lower oxidation state (e.g. $Tl^+$ is more stable than $In^+$ relative to their respective +3 ions, and $Pb^{2+}$ is more stable than $Sn^{2+}$ relative to their respective +4 ions. In the case of thallium, $Tl^+$ is *much* more stable than $Tl^{3+}$, which is a potent oxidizing agent. The following chemical equations are manifestations of these trends:

(i)  $Sn^{2+} (aq) + PbO_2 (s) + 4H^+ (aq) \rightarrow Sn^{4+} (aq) + Pb^{2+} (aq) + 2H_2O (l)$

(ii)  $3Tl^{3+} (aq) + 2Al (s) \rightarrow 3Tl^+ (aq) + 2Al^{3+} (aq)$

(iii)  $3In^+ (aq) \rightarrow In^{3+} (aq) + 2In (s)$

(iv)  $2Sn^{2+} (aq) + O_2 (g) + 4H^+ (aq) \rightarrow 2Sn^{4+} (aq) + 2H_2O (l)$

(v)  $Tl^+ (aq) + O_2 (g) \rightarrow NR$

**11.11 Verify the results of the chemical equations above using $E°$ values from Appendix 4.** The net reactions (i) - (v), above, are composed of two half reactions, which are reproduced below.  For reactions (i) - (iv), the sum of the $E°$ values for the two half reactions is positive, which means that the $\Delta G°$ is negative and that the net reaction is spontaneous; for reaction (v), the net $E°$ is negative:

(i)  $Sn^{2+} (aq) \rightarrow Sn^{4+} (aq) + 2e^-$    $E° = 0.15$ V

$PbO_2 (s) + 4H^+ (aq) + 2e^- \rightarrow Pb^{2+} (aq) + 2H_2O (l)$    $E° = 1.698$ V

net $E°$ for (i) = 1.85 V

(ii)  $Tl^{3+} (aq) + 2e^- \rightarrow Tl^+ (aq)$    $E° = 1.25$ V

$Al (s) \rightarrow Al^{3+} (aq) + 3e^-$    $E° = 1.676$ V

net $E°$ for (ii) = 2.04 V

(iii)  $In^+ (aq) \rightarrow In^{3+} (aq) + 2e^-$    $E° = 0.444$ V

$In^+ (aq) + e^- \rightarrow In (s)$    $E° = -0.126$ V

net $E°$ for (iii) = 0.318 V

(iv)  $Sn^{2+} (aq) \rightarrow Sn^{4+} (aq) + 2e^-$    $E° = 0.15$ V

$O_2 + 4H^+ (aq) + 4e^- \rightarrow 2 H_2O$    $E° = 1.229$ V

net $E°$ for (iv) = 1.38 V

(v)  $Tl^+ (aq) \rightarrow Tl^{3+} (aq) + 2e^-$    $E° = -1.25$ V

$O_2 + 4H^+ (aq) + 4e^- \rightarrow 2 H_2O$    $E° = 1.229$ V

net $E°$ for (iv) = -0.02 V

**11.12 The reactions of $K_2CO_3$ and $Na_4SiO_4$ with acid.** Both of these compounds react with acid to produce the oxide, which for carbon is $CO_2$ and for silicon is $SiO_2$. The balanced equations are:

$$K_2CO_3 \ (aq) \ + \ 2HCl \ (aq) \ \rightarrow \ 2KCl \ (aq) \ + \ CO_2 \ (g) \ + \ H_2O \ (l)$$

$$Na_4SiO_4 \ (aq) \ + \ 4HCl \ (aq) \ \rightarrow \ 4NaCl \ (aq) \ + \ SiO_2 \ (s) \ + \ 2H_2O \ (l)$$

The second equation represents one of the ways that silica gel is produced.

**11.13 The nature of $[SiO_3^{2-}]_n$ ions and layered aluminosilicates.** In contrast with the extended three-dimensional structure of $SiO_2$, the structures of jadeite and kaolinite consist of extended one- and two-dimensional structures, respectively. The $[SiO_3^{2-}]_n$ ions in jadeite are a linear polymer of $SiO_4$ tetrahedra, each one sharing a bridging oxygen atom with the tetrahedron before it and the tetrahedron after it in the chain (this is referred to as a chain metasilicate; see **Structural Drawing 19**). Each silicon atom has two bridging oxygen atoms and two terminal oxygen atoms. The two-dimensional aluminosilicate layers in kaolinite represent another way of connecting $SiO_4$ tetrahedra (see **Figure 11.11**). Each silicon atom has three oxygen atoms that bridge to other silicon atoms in the plane and one oxygen atom that bridges to an aluminum atom.

**11.14 Molecular sieves: (a) How many bridging O atoms are in a single sodalite cage?** A sodalite cage is based on a truncated octahedron. Refer to the answer to **In-Chapter Exercise 11.6**. Each of the heavy lines in the drawing of an octahedron truncated along one of its $C_4$ axes represents an M–O–M linkage (M = Si or Al). There are eight such lines in the drawing, and since the truncation procedure is carried out six times to produce a sodalite cage, there are 8 x 6 = 48 bridging oxygen atoms.

   **(b) Describe the polyhedron at the center of the zeolite A structure.** As shown in **Figure 11.14**, eight sodalite cages are linked together to form the large α cage of zeolite A. The polyhedron at the center has six octagonal faces (the one in the front of the diagram is the most obvious) and eight smaller square faces. If you look at **Figure 11.13(a)**, you will see that a truncated octahedron has 8 hexagonal faces and 4 smaller square faces. Thus, the fusing together of 8 truncated octahedra produces a central cavity that is different than a truncated octahedron.

**11.15  Compare pyrophilite and muscovite mica.** The generic structure of the 2:1 aluminosilicates that include pyrophylite and muscovite mica is shown in **Figure 11.12**. The discrete O–Si–O–Al–O–Si–O layers are composed of a layer of $AlO_6$ octahedral sandwiched between layers of $SiO_4$ tetrahedra. In pyrophylite, which has the formula $Al_2(OH)_2Si_4O_{10}$ (note: 1 Al per 2 Si, hence the name 2:1 aluminosilicate), the O–Si–O–Al–O–Si–O layers are electrically neutral, leading to weak van der Waals bonds between layers despite the strong Al–O and Si–O bonding within each layer. In muscovite mica, which has the formula $KAl_2(OH)_2Si_3AlO_{10}$, one Al(III) ion replaces a Si(IV) ion in the layers of $SiO_4$ tetrahedra (i.e. there are some $AlO_4$ tetrahedra within these layers. The result of this is that the layers are now negatively charged, and the potassium ions compensate the negative charge. They are accomodated between the layers, so the layers in muscovite mica are held together by relatively strong K–O ionic bonds. This accounts for the fact that mica is much harder than talc.

**11.16  Give the structural type and describe the structures of $B_4H_{10}$, $B_5H_9$, and 1,2-$B_{10}C_2H_{12}$.** Simple polyhedral boranes come in three basic types, $B_nH_n{}^{2-}$ *closo* structures, $B_nH_{n+4}$ *nido* structures, and $B_nH_{n+6}$ *arachno* structures. The first compound given in this **Exercise**, $B_4H_{10}$, is an example of a $B_nH_{n+6}$ compound with n = 4, so it is an *arachno* borane. Its structure is shown in **Structural Drawing 23**: Two B–H units are joined by a 2c,2e B–B bond; this $B_2H_2$ unit is flanked by four hydride bridges to two $BH_2$ units (the B–H–B bridge bonds are 3c,2e bonds). The second compound, $B_5H_9$, is an example of a $B_nH_{n+4}$ compound with n = 5, so it is a *nido* borane. Its structure is shown in **Structural Drawing 26**: Four B–H units are joined by four hydride bridges; the resulting $B_4H_8$ unit, in which the four boron atoms are coplanar, is capped by an apical B–H unit that is bonded to all four of the coplanar boron atoms. The third compound is an example of a carborane in which two C–H units substitute for two B–H$^-$ units in $B_{12}H_{12}{}^{2-}$, resulting in 1,2-$B_{10}C_2H_{12}$ that retains the *closo* structure of the parent $B_{12}H_{12}{}^{2-}$ ion. Its structure is shown in **Structural Drawing 32**: A $B_5H_5$ pentagonal plane is joined to a

$B_4CH_5$ pentagonal plane that is offset from the first plane by 36°; The first plane is capped by a B–H unit while the second is capped by a C–H unit.

**11.17  Classify $B_{10}H_{14}$ and discuss its structure and bonding with respect to Wade's rules.**  Refer to the answer to **End-of-Chapter Exercise 11.16**, above. This compound is an example of a $B_nH_{n+4}$ compound with n = 10, so it is a *nido* borane.  According to Wade's rules, ten B–H units contribute 10 x 2 = 20 electrons, and the four additional H atoms contribute four additional electrons (24 electrons = 12 pairs of skeletal electrons).  The structure of $B_{10}H_{14}$ is shown in **Figure 11.15** and in **Table 11.8**.  The total number of valence electrons for $B_{10}H_{14}$ is (10 x 3) + (14 x 1) = 44.  Since there are 10 2c,2e B–H bonds, which account for 20 of the valence electrons, the number of cluster valence electrons is the remainder, 44 – 20 = 24.

**11.18  The synthesis of $Fe(nido\text{-}B_9C_2H_{11})_2$.**  The starting material, $B_{10}H_{14}$, is converted to the *closo* carborane $1,2\text{-}B_{10}C_2H_{12}$ by treatment with acetylene in the presence of a Lewis base, usually diethylsulfide:

$$B_{10}H_{14} + C_2H_2 \xrightarrow{\text{SEt}_2} B_{10}C_2H_{12} + 2H_2$$

This compound is fragmented by the removal of a B atom as $B(OEt)_3$:

$$B_{10}C_2H_{12} + Na^+OEt^- + 2EtOH \rightarrow Na^+[B_9C_2H_{12}]^- + B(OEt)_3 + H_2$$

This salt is deprotonated to the –2 anion and then treated with $FeCl_2$ to form the final product:

$$Na^+[B_9C_2H_{12}]^- + NaH \rightarrow (Na^+)_2[B_9C_2H_{11}]^{2-} + H_2$$

$$2(Na^+)_2[B_9C_2H_{11}]^{2-} + FeCl_2 \rightarrow (Na^+)_2[Fe(B_9C_2H_{11})_2]^{2-} + 2NaCl$$

**11.19  Compare BN and graphite in terms of:  (a)  Their structures.**  Both of these substances have layered structures.  The planar sheets in boron nitride and in graphite consist of edge-shared hexagons such that each B or N atom in BN has three nearest neighbors that are the other type of atom and each C atom in graphite has three nearest neighbor C atoms. The B–N and C–C distances within the sheets, 1.45 Å and 1.42 Å, respectively, are much shorter than the perpendicular interplanar spacing, 3.33 Å and 3.35 Å, respectively. In BN, the $B_3N_3$ hexagonal rings are stacked directly over one another so that B and N atoms from alternating planes are 3.33 Å apart, while in graphite the $C_6$ hexagons are staggered (see **Figure 11.7**) so that C atoms from alternating planes are either 3.35 Å or 3.64 Å apart (you should determine this yourself using trigonometry).

(b) **Their reactivity with Na and Br$_2$.** Graphite reacts with alkali metals and with halogens (see the answer to **In-Chapter Exercise 11.4**). In contrast, boron nitride is quite unreactive. The large HOMO-LUMO gap in BN, which causes it to be an insulator, suggests an explanation for the lack of reactivity: since the HOMO of BN is a relatively low energy orbital, it is more difficult to remove an electron from it than from the HOMO of graphite, and since the LUMO of BN is a relatively high energy orbital, it is more difficult to add an electron to it than to the LUMO of graphite.

**11.20 Devise a synthesis for the following borazines: (a) Ph$_3$N$_3$B$_3$Cl$_3$.** The reaction of a primary ammonium salt with boron trichloride yields N-substituted B-trichloroborazines:

$$3PhNH_3{}^+Cl^- + 3BCl_3 \rightarrow Ph_3N_3B_3Cl_3 + 9HCl$$

(b) **Me$_3$N$_3$B$_3$H$_3$.** We first prepare $Me_3N_3B_3Cl_3$ using $MeNH_3{}^+Cl^-$ and the method described above, and then perform a $Cl^-/H^-$ metathesis reaction using LiH as the hydride source:

$$3MeNH_3{}^+Cl^- + 3BCl_3 \rightarrow Me_3N_3B_3Cl_3 + 9HCl$$

$$Me_3N_3B_3Cl_3 + 3LiH \rightarrow Me_3N_3B_3H_3 + 3LiCl$$

The structures of $Ph_3N_3B_3Cl_3$ and $Me_3N_3B_3H_3$ are shown below:

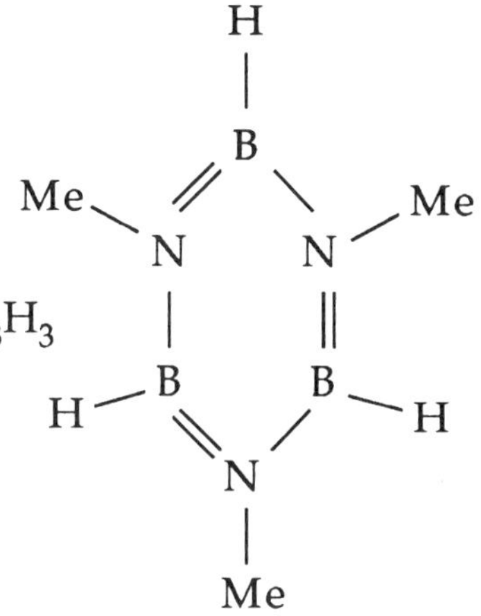

## 11.21 Electrical Conductivities. (a) Describe the trend in $E_g$ for C (diamond) through Sn (gray), and for cubic BN, AlP, and GaAs.

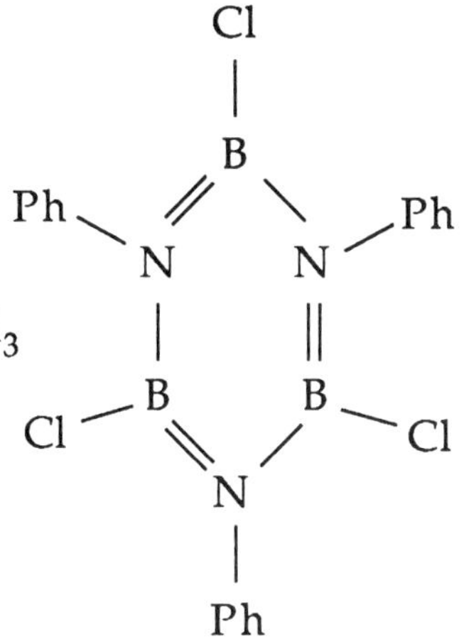

Reference to **Table 11.9** will show that the bandgap decreases considerably from diamond (5.47 eV) to gray tin (~0 eV). In harmony with this trend of decreasing bandgap down a group, the bandgaps of the III-V compounds BN, AlP, and GaAs decrease in that order (both B and N are from Period 2, both Al and P are from Period 3, and both Ga and As are from Period 4).

**(b) Temperature dependence of $\sigma(Si)$.** Silicon is a semiconductor. This type of material always experiences an increase in conductivity as the temperature is raised (see **Section 3.6**). Therefore, the conductivity of Si will be greater at 40 °C than at 20 °C. You should recall that the conductivity of a metal decreases as the temperature is raised.

**(c) Temperature dependence of AlP and GaAs.** The functional dependence of conductivity on temperature is:

$$\sigma = \sigma_0 e^{-E_g/2kT}$$

Accordingly, the greater the bandgap $E_g$, the more $\sigma$ will change for a given change in temperature. Since AlP has a larger bandgap than GaAs, the conductivity of the former compound will be more sensitive to temperature.

# CHAPTER 12

# THE NITROGEN AND OXYGEN GROUPS

| 14 | 15 | 16 | 17 |
|----|----|----|----|
| C | **N** | **O** | F |
| Si | **P** | **S** | Cl |
| Ge | **As** | **Se** | Br |
| Sn | **Sb** | **Te** | I |
| Pb | **Bi** | **Po** | At |

V    VI

The elements of the nitrogen and oxygen groups contain no true metals, unlike the elements of groups 13/III and 14/IV. They form compounds in a wide variety of oxidation states (e.g., $N_2O$, $NO$, and $NO_2$; $S_2O_4^{2-}$, $SO_3^{2-}$, and $SO_4^{2-}$), and therefore have a rich redox chemistry. The heavier members form ring and cluster compounds. Some of the industrially most important inorganic chemicals come from these two groups, including sulfuric acid and ammonia.

## SOLUTIONS TO IN-CHAPTER EXERCISES

**12.1   The structure of bismuth.** Considering the three nearest neighbors only, the structure around each Bi atom is trigonal pyramidal, like $NH_3$. A side view of the structure of bismuth is shown to the right. VSEPR theory predicts this geometry for an atom that has a Lewis structure with three bonding pairs and a lone pair.

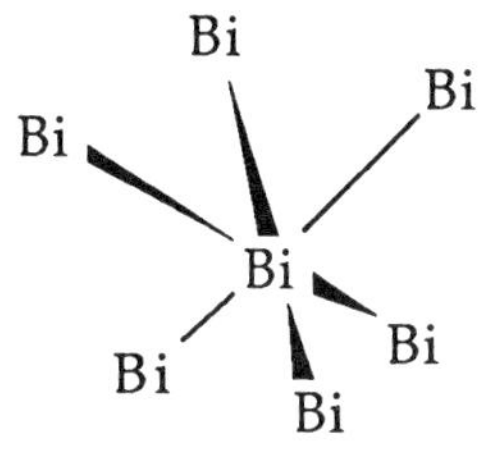

**12.2 Is phosphorus or sulfur a stronger oxidizing agent?** The key to estimating the oxidizing strength of a substance is to determine how readily it is reduced. From the table of standard reduction potentials in **Appendix 4** we can obtain the following information:

$$P\ (s)\ +\ 3\ e^-\ +\ 3\ H^+\ (aq)\ \rightarrow\ PH_3\ (aq) \qquad E^\circ = -0.063\ V$$

$$S\ (s)\ +\ 2\ e^-\ +\ 2\ H^+\ (aq)\ \rightarrow\ H_2S\ (aq) \qquad E^\circ = 0.144$$

Since $\Delta G = -nFE^\circ$, sulfur is more readily reduced and is therefore the better oxidizing agent. This is in harmony with the higher electronegativity of sulfur ($\chi$ = 2.58) than phosphorus ($\chi$ = 2.19). As you shall see in **Chapter 13**, the most electronegative element, fluorine, is also the most potent elemental oxidizing agent.

**12.3 The syntheses of hydrazine and hydroxylamine.** The reactions that are employed to synthesize hydrazine are as follows:

$$NH_3\ +\ ClO^-\ +\ H^+\ \rightarrow\ [\,H_3N\!:\!\rightarrow Cl\!-\!O^-\,]\ \rightarrow\ H_2NCl\ +\ H_2O$$

$$H_2NCl\ +\ NH_3\ \rightarrow\ H_2NNH_2\ +\ HCl$$

Both of these can be thought of as redox reactions since the formal oxidation state of the N atoms changes (from –3 to –1 in the first reaction and from –1 and –3 to –2 and –2 in the second reaction). Mechanistically, both reactions appear to involve nucleophilic attack by $NH_3$ (a Lewis base) on either $ClO^-$ or $NH_2Cl$ (acting as Lewis acids). The reaction employed to synthesize hydroxylamine, which produces the intermediate $N(OH)(SO_3)_2$, probably involves the attack of $HSO_3^-$ (acting as a Lewis base) on $NO_2^-$ (acting as a Lewis acid), although since the formal oxidation state of the N atom changes from +3 in $NO_2^-$ to +1 in $N(OH)(SO_3)_2$ this can also be seen as a redox reaction. Whether one considers reactions such as these to be redox reactions or nucleophilic substitutions may depend on the context in which the reactions are being discussed. It is easy to see that these reactions do not involve simple electron transfer, such as in $2Cu^+\ (aq) \rightarrow Cu^0\ (s)\ +\ Cu^{2+}\ (aq)$.

**12.4 Synthesis of a polyphosphazene high polymer.** Cyclic phosphazene dichlorides such as $(Cl_2PN)_3$ (**Structural Drawing 17**) and $(Cl_2PN)_4$ (**Structural Drawing 18**) are convenient starting materials for polydichlorophosphazene, which can be converted to other polyphosphazenes by nucleophilic substitution of the P–Cl bonds. In this case, we use the $N(CH_3)_2^-$ anion as the nucleophile (its lithium salt can be prepared from dimethylamine and lithium metal):

$$3PCl_5 + 3NH_4Cl \rightarrow (Cl_2PN)_3 + 12HCl$$

$$(Cl_2PN)_3 \xrightarrow{290°} (Cl_2PN)_n$$

$$(Cl_2PN)_n + 2nLiN(CH_3)_2 \rightarrow (((CH_3)_2N)_2PN)_n + 2nLiCl$$

**12.5 Can $Cl^-$ or $Br^-$ catalyze the decomposition (disproportionation) of hydrogen peroxide?** We must determine is the standard potentials for the reduction of $Cl_2$ to $Cl^-$ or $Br_2$ to $Br^-$ are within the range bounded by the reduction of $O_2$ to $H_2O_2$ (0.695 V) and the reduction of $H_2O_2$ to $H_2O$ (1.76 V). From **Appendix 4** we can get the following information:

$$Cl_2\ (aq) + 2e^- \rightarrow 2Cl^-\ (aq) \qquad E° = 1.358\ V$$

$$Br_2\ (aq) + 2e^- \rightarrow 2Br^-\ (aq) \qquad E° = 1.087\ V$$

Since both of the reduction potentials for these reactions are within the prescribed range, both chloride and bromide can catalyze the decomposition of $H_2O_2$. Great care must be exercised when storing $H_2O_2$ for long periods of time, since trace amounts of these common anions will slowly destroy the compound.

**12.6 Structures of $SO_3$ and $SO_3F^-$.** The Lewis structures are shown below (only one of the three possible resonance structures for $SO_3$ are shown). The central S atom in sulfur trioxide is trigonal planar, and since all of the S–O bonds are equivalent, the point group is $D_{3h}$. The central S atom in fluorosulfate ion is tetrahedral, but the ion is $C_{3v}$, not $T_d$.

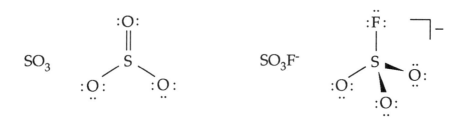

**12.7   Why is $MoS_2$ an effective lubricant.**   The layered structure of molybdenum(IV) sulfide is shown in **Figure 12.8**.  Although the Mo–S bonds are undoubtedly quite strong, the S⋯S interactions between adjacent S–Mo–S layers are weak and easily disrupted (these are best thought of as van der Waals interactions).  The slipperiness of $MoS_2$ is due to the ease with which one layer can glide over another.

**12.8   The structure of $Ge_9{}^{2-}$.**   Each of the germanium atoms in this cluster has four valence electrons, so the cluster has $(9 \times 4) + 2 = 38$ electrons or 19 electron pairs.  If we assume that each atom has a lone pair that is directed away from the other atoms in the cluster (and hence is unavailable for Ge–Ge framework bonding), there are 10 pairs of skeletal electrons, or one more than the number of cluster atoms.  This suggests a *closo*  structure, and the structure of $Ge_9{}^{2-}$ shown in **Structural Drawing 37** has the same framework structure as the *closo* structure of $B_9H_9{}^{2-}$ shown in **Figure 11.15**.

SOLUTIONS TO END-OF-CHAPTER EXERCISES

**12.1 General properties of the elements.** In the list below, the properties of polonium have been omitted since the chemistry of this element has been little studied owing to its radioactivity.

|  | type of element | diatomic gas? | achieves maximum oxidation state? | displays inert pair effect? |
|---|---|---|---|---|
| N | nonmetal | yes | yes | no |
| P | nonmetal | no | yes | no |
| As | nonmetal | no | yes | no |
| Sb | nonmetal | no | yes | no |
| Bi | metalloid | no | yes | yes |
| O | nonmetal | yes | no | no |
| S | nonmetal | no | yes | no |
| Se | nonmetal | no | yes | no |
| Te | nonmetal | no | yes | no |

**12.2 Contrast the formulas and stabilities of: (a) Nitrogen and phosphorus chlorides.** The only isolable nitrogen chloride is $NCl_3$, and it is thermodynamically unstable with respect to its constituent elements (i.e. it is endoergic). The compound $NCl_5$ is unknown. In contrast, both $PCl_3$ and $PCl_5$ are stable and can be prepared directly from phosphorus and chlorine.

**(b) Oxygen and sulfur fluorides.** Oxygen forms two binary fluorides, $O_2F_2$, which is endoergic, and $OF_2$, which is exoergic. Sulfur forms $S_2F_2$, which is homologous to $O_2F_2$, as well as $SF_4$, $S_2F_{10}$, and $SF_6$, and all of these are exoergic. Note that sulfur does not form $SF_2$.

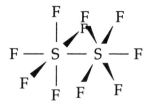

The staggered structure of $S_2F_{10}$, which has $D_{4d}$ symmetry.

**12.3   The probable structures of: (a) SeF₄.** The Lewis structure of this compound is shown below. VSEPR theory predicts that a compound with four bonding pairs of electrons and one lone pair will exhibit a $C_{2v}$ saw-horse structure (see **Structural Drawing 4** in **Chapter 2**, which shows the structure of the isostructural molecule SF₄).

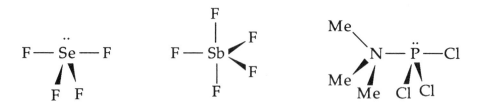

**(b)   SbF₅ (g).** The Lewis structure of this molecule is shown above. VSEPR theory predicts that a compound with five bonding pairs of electrons will exhibit a $D_{3h}$ trigonal bipyramidal structure.

**(c)   The N(CH₃)₃ complex of PCl₃.** The Lewis structure of this complex, which is the same as the one for SeF₄, is also shown above. The saw–horse structure has lower symmetry than SeF₄, since the N(CH₃)₃ group is unique. It is generally found that groups that wind up in the trigonal bipyramidal axial positions are the ones that are the most electronegative, so it is no surprise that the trimethylamine group is axial. The complex has $C_s$ symmetry.

**12.4   Describe the relative oxidizing strengths and Lewis basicities of isoelectronic species.** Of the species listed, $O_2^{2-}$ and $N_2H_3^-$ are isoelectronic and $C_2^{2-}$, $CN^-$, $N_2$, and $NO^+$ are isoelectronic. Considering the former pair, the species with the higher negative charge ($O_2^{2-}$) is the stronger Lewis base, and the species with the lower negative charge ($N_2H_3^-$) is more easily reduced and is therefore the stronger oxidant. Considering the latter group, $NO^+$ will be the most easily reduced and therefore will be the strongest oxidant. Accordingly, it will also be the weakest Lewis base. Once again basicity increases and oxidizing strength decreases with increasing negative charge.

**12.5**   **Give the formula and name of a species that is isoelectronic and isostructural with: (a)** $NO_3^-$. Two examples are carbonate dianion, $CO_3^{2-}$, and the gaseous form of sulfur trioxide, $SO_3$.

   **(b)** $NO_2^-$. Two examples of isoelectronic and isostructural species are ozone, $O_3$, and sulfur dioxide, $SO_2$.

   **(c)** $N_2O_4$. An isoelectronic and isostructural species is oxalate dianion, $C_2O_4^{2-}$.

   **(d)** $N_2O$. An isoelectronic and isostructural species is carbon dioxide, $CO_2$.

   **(e)** $N_2$. Two examples are cyanide ion, $CN^-$, and nitrosyl ion, $NO^+$.

   **(f)** $NH_3$. Two examples are methyl carbanion, $CH_3^-$, and oxonium ion (or hydronium ion), $H_3O^+$.

**12.6**   **Contrast acidity and oxidizing strength of isoelectronic species: (a)** $CO_3^{2-}$, $NO_3^-$, **and** $SO_3$. As explained in the answer to **End-of-Chapter Exercise 12.4**, in general the species with the higher negative charge is the stronger Lewis base and hence the weaker Lewis acid, and the species with the lower negative charge is more easily reduced and is therefore the stronger oxidant. However, the oxidizing strength also depends on the position of the central element in its respective group. Therefore, although $SO_3$ is the strongest Lewis acid of the three species, it is not the strongest oxidant. Nitrate is a stronger oxidant than sulfur trioxide.

(b) $NO_2^-$, $O_3$, and $SO_2$. Sulfur dioxide is the strongest Lewis acid of the three species. Ozone is the strongest oxidant.

(c) $C_2O_4^{2-}$ and $N_2O_4$. Dinitrogen tetroxide is the stronger Lewis acid and the stronger oxidant.

(d) $N_2O$ and $CO_2$. Carbon dioxide is the stronger Lewis acid (carbon is less electronegative than nitrogen). Nitrous oxide is the stronger oxidant (the product of the reduction of $N_2O$ is a species with an enormously stable triple bond, $N_2$).

(e) $CN^-$, $N_2$, and $NO^+$. The strongest Lewis acid and oxidant is $NO^+$.

(f) $CH_3^-$, $NH_3$, and $H_3O^+$. None of these species exhibit Lewis acidity at the central atom, but the latter two can participate in hydrogen bonding. Due to its positive charge, $H_3O^+$ will hydrogen bond more strongly than $NH_3$. The hydronium ion is also the strongest oxidant.

**12.7   Give equations and conditions for the synthesis of: (a) $HNO_3$.** The synthesis of nitric acid, an example of the most oxidized form of nitrogen, starts with ammonia, the most reduced form:

$$4NH_3 \,(aq) + 7O_2 \,(g) \rightarrow 6H_2O \,(g) + 4NO_2 \,(g)$$

High temperatures are obviously necessary for this reaction to proceed at a reasonable rate — ammonia is a flammable gas but at room temperature it does not react rapidly with air. The second step in the formation of nitric acid is the high–temperature disproportionation of $NO_2$ in water:

$$3NO_2 \,(aq) + H_2O \,(l) \rightarrow 2HNO_3 \,(aq) + NO \,(g)$$

(b) $NO_2^-$. Whereas the disproportionation of $NO_2$ in acidic solution yields $NO_3^-$ and $NO$ (see part (a), above), in basic solution nitrite ion is formed:

$$2NO_2 \,(aq) + 2OH^- \,(aq) \rightarrow NO_2^- \,(aq) + NO_3^- \,(aq) + H_2O \,(l)$$

   **(c) NH$_2$OH.**  The protonated form of hydroxylamine is formed in a very unusual reaction between nitrite ion and bisulfite ion in cold aqueous acidic solution:

$$NO_2^- \ (aq) \ + \ 2HSO_3^- \ (aq) \ + \ H_2O \ (l) \ \rightarrow \ NH_3OH^+ \ (aq) \ + \ 2SO_4^{2-} \ (aq)$$

The $NH_3OH^+$ ion can be deprotonated with base.

   **(d) N$_3^-$.**  The azide ion can be prepared from anhydrous molten sodium amide (m.p. ~200 °C) and either nitrate ion or nitrous oxide at elevated temperatures:

$$3NaNH_2 \ (l) \ + \ NaNO_3 \ \rightarrow \ NaN_3 \ + \ 3NaOH \ + \ NH_3 \ (g)$$

$$2NaNH_2 \ (l) \ + \ N_2O \ \rightarrow \ NaN_3 \ + \ NaOH \ + \ NH_3$$

**12.8   The balanced chemical equation for the formation of P$_4$O$_{10}$ (s).**  This compound is formed by the complete combustion of elemental phosphorus, as shown below:

$$P_4 \ (s) \ + \ 5O_2 \ (g) \ \rightarrow \ P_4O_{10} \ (s)$$

As discussed in **Section 12.1**, white phosphorus (P$_4$) is adopted as the reference phase for thermodynamic calculations even though it is not the most stable phase of elemental phosphorus.

**12.9   Frost diagrams for phosphorus and bismuth.**  The main points you will want to remember are  (i)  Bi(III) is much more stable than Bi(V) and  (ii)  P(III) and P(V) are both about equally stable (i.e.,  Bi(V) is a strong oxidant but P(V) is not).  This suggests that the point for Bi(III) on the Frost diagram lies *below* the line connecting Bi(0) and B(V), while the point for P(III) lies very close to the line connecting P(0) and P(V).  The essential parts of the Frost diagrams for these two elements are shown at the right.  The complete diagrams are shown in **Figure 12.5**.

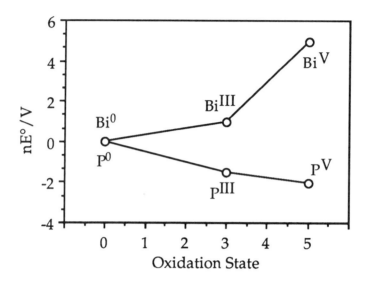

**12.10  The pH dependence of the rate of reduction of $NO_2^-$.** The rates of reactions in which nitrite ion is reduced (i.e. in which it acts as an oxidizing agent) are increased as the pH is lowered.  That is, acid enhances the rate of oxidations by $NO_2^-$.  The reason is that $NO_2^-$ is converted to the nitrosonium ion, $NO^+$, in strong acid:

$$HNO_2 \ (aq) \ + \ H^+ \ (aq) \ \rightarrow \ NO^+ \ (aq) \ + \ H_2O \ (l)$$

This cationic Lewis acid can form complexes with the Lewis bases undergoing oxidation (species that are oxidizable are frequently electron rich and hence are basic).  Therefore, at low pH the oxidant ($NO^+$) is a different chemical species than at higher pH ($NO_2^-$).

**12.11  The reaction of $H_3PO_2$ with $Cu^{2+}$.** From the standard potentials for the following two half-reactions, the standard potential for the net reaction can be calculated:

$$H_3PO_2 \ (aq) \ + \ H_2O \ (l) \ \rightarrow \ H_3PO_3 \ (aq) \ + \ 2e^- \ + \ 2H^+ \ (aq) \qquad E° = 0.499 \ V$$

$$Cu^{2+} \ (aq) \ + \ 2e^- \ \rightarrow \ Cu^0 \ (s) \qquad E° = 0.340 \ V$$

Therefore the net reaction is:

$$H_3PO_2 \ (aq) \ + \ H_2O \ (l) \ + \ Cu^{2+} \ (aq) \ \rightarrow \ H_3PO_3 \ (aq) \ + \ Cu^0 \ (s) \ + \ 2H^+ \ (aq)$$

and the net standard potential is $E° = 0.839$ V. To determine whether $HPO_3^{2-}$ and $H_2PO_2^-$ are useful as oxidizing or reducing agents, we must compare $E$ values for their oxidations and reductions at pH = 14 (these potentials are listed in the Basic Solution Latimer diagrams in **Appendix 4**):

$$HPO_3^{2-} \ (aq) + 3OH^- \ (aq) \rightarrow PO_4^{3-} \ (aq) + 2e^- + 2H_2O \ (l) \quad E(pH = 14) = 1.12 \ V$$

$$HPO_3^{2-} \ (aq) + 2e^- + 2H_2O \ (l) \rightarrow H_2PO_2^- \ (aq) + 3OH^- \ (aq) \quad E(pH = 14) = -1.57 \ V$$

$$H_2PO_2^- \ (aq) \ + \ e^- \ \rightarrow \ P \ (s) \ + \ 2OH^- \ (aq) \quad E(pH = 14) = -2.05 \ V$$

Since a positive potential will give a negative value of $\Delta G$, the oxidations of $HPO_3^{2-}$ and $H_2PO_2^-$ are much more favorable than their reductions, so these ions will be much better reducing agents than oxidizing agents.


**12.12  Disproportionations: (a)  The disproportionation of $H_2O_2$ and $HO_2$.** To calculate the standard potential for a disproportionation reaction, you must sum the potentials for the oxidation and reduction of the species in question.   For hydrogen peroxide in acid solution, the oxidation and reduction are:

$$H_2O_2 \ (aq) \ \rightarrow \ O_2 \ (g) \ + \ 2e^- \ + \ 2H^+ \ (aq) \quad E° = -0.695 \ V$$

$$H_2O_2 \ (aq) \ + \ 2e^- \ + \ 2H^+ \ (aq) \ \rightarrow \ 2H_2O \ (l) \quad E° = 1.763 \ V$$

Therefore, the standard potential for the net reaction $2H_2O_2 \ (aq) \rightarrow O_2 \ (g) + 2H_2O \ (l)$ is $(-0.695 \ V) + (1.763 \ V) = 1.068 \ V$.

     **(b)  Catalysis by $Cr^{2+}$?** As discussed in **In-Chapter Exercise 12.5**, $Cr^{2+}$ can act as a catalyst for the decomposition of hydrogen peroxide if the $Cr^{3+}/Cr^{2+}$ reduction potential falls between the values for the reduction of $O_2$ to $H_2O_2$ (0.695 V) and the reduction of $H_2O_2$ to $H_2O$ (1.76 V). Reference to

**Appendix 4** reveals that the $Cr^{3+}/Cr^{2+}$ reduction potential is $-0.424$ V, so $Cr^{2+}$ is *not* capable of decomposing $H_2O_2$.

(c) **The disproportionation of $HO_2$.** The oxidation and reduction of superoxide ion $(O_2^-)$ in acid solution are:

$$HO_2\ (aq)\ \rightarrow\ O_2\ (g)\ +\ e^-\ +\ H^+\ (aq) \qquad E° = 0.125\ V$$

$$HO_2\ (aq)\ +\ e^-\ +\ H^+\ (aq) \qquad E° = 1.51\ V$$

Therefore, the standard potential for the net reaction $2HO_2\ (aq)\ \rightarrow\ O_2\ (g)\ +\ H_2O_2\ (aq)$ is $(0.125\ V) + (1.51\ V) = 1.63$ V. Since 1 V is equivalent to 96.5 kJ mol$^{-1}$, $\Delta G° = 157$ kJ mol$^{-1}$. For the disproportionation of $H_2O_2$ (part (a)), $\Delta G° = 103$ kJ mol$^{-1}$.

**12.13  Is ethylenediamine or sulfur dioxide a better solvent for $Na_2S_4$, $K_2Te_3$, and $Cd_2(Al_2Cl_7)$?** As explained in the answer to **End-of-Chapter Exercise 6.12**, anionic species such as $S_4^{2-}$ and $Te_3^{2-}$ are intrinsically basic species that cannot be studied in solvents that are Lewis acids because complex formation will destroy the independent identity of the anion. Therefore, since the basic solvent ethylenediamine will not react with $Na_2S_4$ or with $K_2Te_3$, it is a better solvent for them than sulfur dioxide. On the other hand, the $Al_2Cl_7^-$ anion is not basic, despite its negative charge. This is because it can readily dissociate into two species, one of which is strongly acidic ($AlCl_3$) and the other of which is only weakly basic ($AlCl_4^-$). In order to inhibit the formation of an acid-base complex with the strong Lewis acid $AlCl_3$, we should choose a solvent that is itself a Lewis acid, in this case $SO_2$.

**12.14  The $d$-metal monoxides and monosulfides.** The rock-salt (NaCl) structure (see **Figure 4.11**) is exhibited by most of the $d$-metal monoxides, in harmony with the expectation that ionic bonding would predominate with the hard $O^{2-}$ anion in combination with +2 cations. In contrast, most $d$-metal monosulfides have the NiAs structure (see **Figure 4.16**). This structure is generally exhibited when covalent bonding plays an equal or predominant role relative to ionic bonding, exactly the situation that exists for compounds of the soft $S^{2-}$ anion. The reason that monoxides are not common for the

Period 5 and 6 $d$-block elements is that these elements tend to form compounds in their higher or highest oxidation states. In contrast to the $p$-block elements, for which the maximum oxidation state of a group decreases towards the bottom of the group (the inert pair effect), the maximum oxidation state of the $d$-block elements increases down a group.

**12.15 Explain why NO at low concentrations in automobile exhaust reacts slowly with oxygen (air).** The given observations suggest that more than one NO molecule is involved in the activated complex (i.e. in the rate determining step), since the rate of reaction is dependent on the concentration of NO. Therefore, the rate law must be more than first order in NO concentration. It turns out to be second order: an equilibrium between NO and its dimer $N_2O_2$ precedes the rate determining reaction with oxygen. At high concentrations of NO, the concentration of the dimer is higher.

**12.16 Use isoelectronic analogies to infer the probable structures of: (a) [Sb$_4$]$^{2-}$.** Antimony is in Group 15/V, as is Bismuth. Therefore, $[Sb_4]^{2-}$ is isoelectronic with $[Bi_4]^{2-}$, which has the planar $D_{4h}$ structure shown in **Structural Drawing 36.** Both of these ions have 22 valence electrons ((5 x 4) + 2 = 22). Other $p$-block clusters with four atoms and 22 valence electrons are $Se_4^{2+}$ and $Te_4^{2+}$. Since these are also isoelectronic with $Sb_4^{2-}$ and $Bi_4^{2-}$, and consistent with this, they also have planar $D_{4h}$ structures (see **Structural Drawing 38**).

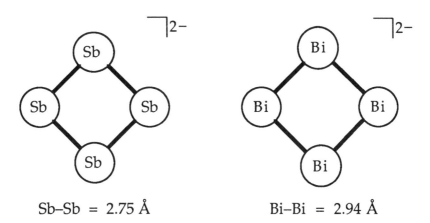

Sb–Sb = 2.75 Å          Bi–Bi = 2.94 Å

**(b) $[P_7]^{3-}$.** This ion has seven atoms and 38 valence electrons (($7 \times 5$) + $3 = 38$). The compound $P_4S_3$, shown in **Structural Drawing 40**, also has seven atoms and 38 valence electrons (($4 \times 5$) + ($3 \times 6$) = 38). Therefore, since isoelectronic species frequently have the same structure, the probable structure of the $[P_7]^{3-}$ ion is the structure exhibited by $P_4S_3$. The compound $Sr_3P_{14}$ does contain discrete $[P_7]^{3-}$ ions with the structure shown below.

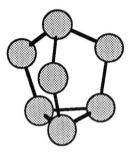

The structure of the $[P_7]^{3-}$ ions in $Sr_3P_{14}$

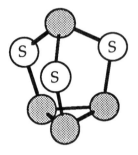

The structure of $P_4S_3$
(see **Structural Drawing 40**)

# CHAPTER 13

# THE HALOGENS AND NOBLE GASES

| | | 18 |
|---|---|---|
| 16 | 17 | **He** |
| O | F | **Ne** |
| S | Cl | **Ar** |
| Se | Br | **Kr** |
| Te | I | **Xe** |
| Po | At | **Rn** |
| VII | VIII | |

The halogens and noble gases are two groups that contain the most reactive and the least reactive elements, respectively. Nevertheless, there are similarities in the patterns of their reactions and in the structures of their compounds. For example, iodine and xenon form an extensive set of iso-electronic and isostructural species, including $IF/XeF^+$, $IF_4^-/XeF_4$, and $IO_3^-/XeO_3$. The halogens form a very important class of compounds with the $d$-block metals, which range from metallic to ionic to molecular. Anhydrous metal halides are among the most important starting materials for inorganic syntheses.

---

## SOLUTIONS TO IN-CHAPTER EXERCISES

**13.1    The structure and point group of $ClO_2F$.** As you should know by now, a reliable way to predict the structure of a $p$-block compound is to draw its Lewis structure and then apply VSEPR theory. The Lewis structure for $ClO_2F$

is shown below. (Note that chlorine is the central atom, not oxygen or fluorine: the heavier, less electronegative element is always the central atom in interhalogen compounds and in compounds containing two different halogens and oxygen.) Three bonding pairs and one lone pair yield a trigonal pyramidal geometry, also shown below. This structure possesses only a single symmetry element, a mirror plane that bisects the O–Cl–O angle and contains the Cl and F atoms. Therefore, $ClO_2F$ has $C_s$ symmetry.

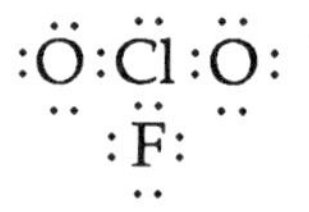

:Ö:Cl̈:Ö:  The Lewis structure        The trigonal pyramidal
   :F̈:      of $ClO_2F$                  structure of $ClO_2F$

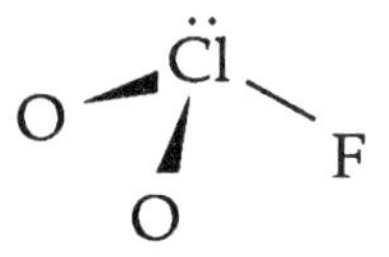

**13.2   Calculate the molar concentration of $IO_4^-$ ions.** As in the **Example**, we can calculate $[H_4IO_6^-]$ using the expression for $K_{a1}$, given that $[H_5IO_6] = 0.30$ M and $[H^+] = 0.70$ M:

$$[H_4IO_6^-] = (0.30\ M/0.70\ M)(5.1 \times 10^{-4}\ M) = 2.2 \times 10^{-4}\ M$$

so therefore,

$$[IO_4^-] = (40)(2.2 \times 10^{-4}\ M) = 8.8 \times 10^{-3}\ M$$

**13.3   Which halogens can oxidize $H_2O$ to $O_2$?** The standard reduction potential for the $O_2/H_2O$ couple is 1.229 V. Any halogen for which the standard reduction potential for the the $X_2/X^-$ couple is more positive than this will oxidize water, since the net potential for the reaction

$$2H_2O\ (aq) + 2X_2\ (aq) \rightarrow O_2\ (g) + 4X^-\ (aq) + 4H^+\ (aq)$$

would then be positive. The $X_2/X^-$ potentials are 3.053 V for fluorine, 1.358 V for chlorine, 1.087 V for bromine, and 0.535 V for iodine. Therefore, elemental fluorine and chlorine are capable of oxidizing water at low pH. However, only the reaction with $F_2$ is rapid. The reaction of $Cl_2$ with $H_2O$ is so slow that commercial production of $Cl_2$ occurs in the presence of $H_2O$ (see **Section 13.1**).

**13.4    The general structural type of: (a) ReF$_6$.** The hexafluorides that are known for Period 4 and 5 $d$-block metals are molecular and generally quite volatile. Rhenium(VI) fluoride is no exception: its boiling point is 34 °C.

**(b)    CoF$_2$.** The halides of Period 3 $d$-block metals in +2 and +3 oxidation states are low-volatility ionic compounds with isolated $M^{2+}$ or $M^{3+}$ ions. The fluorides, including CoF$_2$, generally adopt the rutile structure (see **Table 13.8**).

**(c)    WCl$_6$.** Like the hexafluorides, the two known $d$-block hexachlorides, WCl$_6$ and ReCl$_6$ are volatile, molecular compounds. They are not as volatile as the corresponding fluorides, as shown below for WF$_6$ and WCl$_6$.

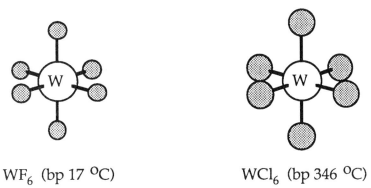

WF$_6$ (bp 17 °C)          WCl$_6$ (bp 346 °C)

**13.5    Suggest a conversion of CoCl$_2$ to CoF$_3$.** Toward the right end of the $d$-block, higher oxidation states become progressively less stable. The compound CoF$_3$ is the only known trihalide of cobalt, and it is a powerful oxidizing agent. Accordingly, it must be prepared using an even more powerful oxidizing agent, such as F$_2$ or ClF$_3$. The usual synthesis uses F$_2$ at 250 °C:

$$2CoCl_2 \,(s) \; + \; 3F_2 \,(g) \; \rightarrow \; 2CoF_3 \,(s) \; + \; 3Cl_2 \,(g)$$

You should appreciate that if excess F$_2$ is present, a mixture of ClF, ClF$_3$, and ClF$_5$ will be formed instead of Cl$_2$. Also, note the similarity between this reaction and the one between Co$_3$O$_4$ and ClF$_3$ discussed in **Section 13.3**.

**13.6 A balanced equation for the decomposition of xenate ion.** We are told that xenate ($HXeO_4^-$) decomposes to perxenate ($XeO_6^{4-}$), xenon, and oxygen, so the equation we must balance is:

$$HXeO_4^- \rightarrow XeO_6^{4-} + Xe + O_2 \quad \text{(not balanced)}$$

Since the reaction occurs in basic solution, we can use $OH^-$ and $H_2O$ to balance the equation. We notice immediately that there is no species containing hydrogen on the right hand side of the equation, so obviously $H_2O$ will go on the right and $OH^-$ will go on the left. The balanced equation is:

$$2HXeO_4^- \, (aq) + 2OH^- \, (aq) \rightarrow XeO_6^{4-} \, (aq) + Xe \, (g) + O_2 \, (g) + 2H_2O \, (l)$$

Since the products are perxenate (a Xe(VIII) species) and elemental xenon, this reaction is a disproportionation of the Xe(VI) species $HXeO_4^-$. Oxygen is produced from a thermodynamically unstable intermediate of Xe(IV), possibly as follows:

$$HXeO_3 \, (aq) \rightarrow Xe \, (g) + O_2 \, (g) + OH^- \, (aq)$$

## SOLUTIONS TO END-OF-CHAPTER EXERCISES

**13.1 General properties of the elements.**

|  | physical state | electronegativity | hardness of halide ion | color |
|---|---|---|---|---|
| $F_2$ | gas | highest (4.0) | hardest | colorless |
| $Cl_2$ | gas | lower | softer | yellow-green |
| $Br_2$ | liquid | lower | softer | dark red-brown |
| $I_2$ | solid | lowest | softest | dark violet |
| He | gas |  |  | colorless |
| Ne | gas |  |  | colorless |
| Ar | gas |  |  | colorless |
| Kr | gas |  |  | colorless |
| Xe | gas |  |  | colorless |

**13.2   Recovery of the halogens from naturally occuring halides.** The principal source of fluorine is $CaF_2$. It is converted to HF by treating it with a strong acid such as sulfuric acid. Liquid HF is electrolyzed to $H_2$ and $F_2$ in an anhydrous cell, with KF as the electrolyte:

$$CaF_2 + H_2SO_4 \rightarrow CaSO_4 + 2HF$$

$$2HF + 2KF \rightarrow 2K^+HF_2^-$$

$$2K^+HF_2^- + electricity \rightarrow F_2 + H_2 + 2KF$$

The principal source of the other halides is sea water (natural brines). Chlorine is liberated by the electrolysis of aqueous NaCl (the chloralkali process):

$$2Cl^- + 2H_2O + electricity \rightarrow Cl_2 + H_2 + 2OH^-$$

Bromine and iodine are prepared by treating aqueous solutions of the halides with chlorine:

$$2X^- + Cl_2 \rightarrow X_2 + 2Cl^- \quad (X^- = Br^-, I^-)$$

**13.3   The chloralkali cell.** A drawing of the cell is shown in **Figure 13.6**. Note that $Cl_2$ is liberated at the anode and $H_2$ is liberated at the cathode, according to the following half reactions:

$$anode: \quad 2Cl^- \,(aq) \rightarrow Cl_2 \,(g) + 2e^-$$

$$cathode: \quad 2H_2O \,(l) + 2e^- \rightarrow 2OH^- \,(aq) + H_2 \,(g)$$

The maintain electroneutrality, $Na^+$ ions diffuse through the polymeric membrane. Due to the chemical properties of the membrane, anions such as $Cl^-$ and $OH^-$ cannot diffuse through it. If $OH^-$ did diffuse through the membrane, it would react with $Cl_2$ and spoil the yield of the electrolysis, as shown in the following equation:

$$2OH^- \,(aq) + Cl_2 \,(aq) \rightarrow ClO^- \,(aq) + Cl^- \,(aq) + H_2O \,(l)$$

**13.4   Sketch the vacant σ\* orbital of a halogen molecule.** According to **Figure 13.1**, the vacant antibonding orbital of a halogen molecule is $2\sigma_u^*$ and is composed primarily of halogen atomic $p$ orbitals (recall that the $ns$-$np$ gap is larger toward the right hand side of the periodic table, and the larger the gap, the smaller the amount of $s$-$p$ mixing). Sketches of the filled $2\sigma_g$ bonding orbital and the empty $2\sigma_u^*$ antibonding orbital are shown below:

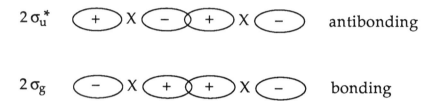

Since the $2\sigma_u^*$ antibonding orbital is the LUMO for a $X_2$ molecule, it is the orbital that accepts the pair of electrons from a Lewis base B when a dative B→$X_2$ bond is formed. From the shape of the LUMO, the B–X–X unit should be linear.

**13.5   The pH dependence of reduction potentials of oxoanions.** The balanced equation for the reduction of $ClO_4^-$ is

$$ClO_4^- \, (aq) \; + \; 2H^+ \, (aq) \; + \; 2e^- \; \rightarrow \; ClO_3^- \, (aq) \; + \; H_2O \, (l)$$

The value of $E°$ for this reaction is 1.201 V (see **Appendix 4**). The potential at any $[H^+]$, given by the Nernst equation, is

$$E \; = \; E° \; - \; ((0.059 \, V/2))(\log([ClO_3^-]/[ClO_4^-][H^+]^2)$$

At pH = 7, $[H^+] = 10^{-7}$ M. Assuming that both perchlorate and chlorate ions are present at unit activity, at pH = 7 the reduction potential is

$$E \; = \; 1.201 \, V \; - \; (0.0295 \, V)(\log 10^{14}) \; = \; 1.201 \, V \; - \; 0.413 \, V \; = \; 0.788 \, V$$

**13.6 Explain why the disproportionation of an oxoanion is promoted by low pH.** Let's take $ClO_3^-$ as an example. Its disproportionation can be broken down into a reduction and an oxidation, as follows:

$$ClO_3^- \ (aq) + 6H^+ \ (aq) + 6e^- \rightarrow Cl^- \ (aq) + 3H_2O \ (l)$$

$$3ClO_3^- \ (aq) + 3H_2O \ (l) \rightarrow 3ClO_4^- \ (aq) + 6H^+ \ (aq) + 6e^-$$

Any effect that changing the pH has on the potential for the reduction reaction will be counteracted by an equal but opposite change on the potential for the oxidation reaction. In other words, the net reaction does not include $H^+$, so the net potential for the disproportionation cannot be pH dependent. Therefore, the promotion of disproportionation reactions of some oxoanions at low pH *cannot* be a thermodynamic promotion. Low pH results in a kinetic promotion — protonation of an oxo group aids oxygen-halogen bond scission (see **Section 13.7**). The disproportionation reactions have the same driving force at high pH and at low pH, but they are much faster at low pH.

**13.7 The analogy between halogens and pseudohalogens. (a) The reaction of NCCN with NaOH.** When a halogen such as chlorine is treated with aqueous base, it undergoes disproportionation to yield chloride ion and hypochlorite ion, as follows:

$$Cl_2 \ (aq) + 2OH^- \ (aq) \rightarrow Cl^- \ (aq) + ClO^- \ (aq) + H_2O \ (l)$$

The analogous reaction of cyanogen with base is:

$$NCCN \ (aq) + 2OH^- \ (aq) \rightarrow CN^- \ (aq) + NCO^- \ (aq) + H_2O \ (l)$$

The linear $NCO^-$ ion is named cyanate.

**(b) The reaction of $SCN^-$ with $MnO_2$ in aqueous acid.** If $MnO_2$ is an oxidizing agent, then the probable reaction is oxidation of thiocyanate ion to thiocyanogen, $(SCN)_2$, coupled with reduction to $MnO_2$ to $Mn^{2+}$ (if you do not recall that $Mn^{3+}$ is unstable to disproportionation, you will discover it when you refer to the Latimer diagram for manganese in **Appendix 4**). The balanced equation is:

$$2SCN^- (aq) + MnO_2 (s) + 4H^+ (aq) \rightarrow (SCN)_2 (aq) + Mn^{2+} (aq) + 2H_2O (l)$$

**(c) The structure of trimethylsilyl cyanide.** Just as halides form Si–X single bonds, trimethylsilyl cyanide contains a Si–CN single bond. Its structure is shown in **Structural Drawing 1**.

**13.8   Iodine fluorides. (a) The structures of $[IF_6]^+$ and $IF_7$.** The structures are shown below.   The structures actually represent the Lewis structures as well, except that the three lone pairs of electrons on each fluorine atom have been omitted.   VSEPR theory predicts that a species with six bonding pairs of electrons, like $IF_6^+$, should be octahedral.   The structures of species with seven bonding pairs of electrons, like $IF_7$, was not covered in **Section 2.2**, but a reasonable and symmetrical structure would be a pentagonal bipyramid.

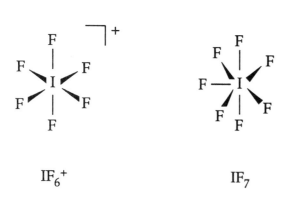

$IF_6^+$                                          $IF_7$

**(b) The preparation of $[IF_6][SbF_6]$.** Judging from the structures shown above, it should be possible to abstract a $F^-$ ion from $IF_7$ to produce $IF_6^+$. The strong Lewis acid $SbF_5$ should be used for the fluoride abstraction so that the salt $[IF_6][SbF_6]$ will result:

$$IF_7 + SbF_5 \rightarrow [IF_6][SbF_6]$$

**13.9   The reaction of $[NR_4][Br_3]$ with $I_2$.** Since $Br_3^-$ has only a moderate formation constant, we can think of it as being chemically equivalent to $Br_2$ + $Br^-$. Since IBr is fairly stable, the probable reaction will be:

$$Br_3^- + I_2 \rightarrow 2IBr + Br^-$$

The IBr may associate to some extent with $Br^-$ to produce $IBr_2^-$.

**13.10 Explain why $CsI_3$ is stable but $NaI_3$ is not.** This is an example of an important principal of inorganic chemistry: large cations stabilize large, unstable anions. This trend appears to have a simple electrostatic origin, as described in **Section 4.8**. A detailed analysis of this particular situation is as follows. The enthalpy change for the reaction:

$$NaI_3 \ (s) \ \rightarrow \ NaI \ (s) \ + \ I_2 \ (s)$$

is negative, which is just another way of stating that $NaI_3$ is not stable. This enthalpy change is composed of four terms, as shown below:

$$\Delta H \quad = \quad \begin{array}{l} \text{lattice enthalpy of } NaI_3 \\ + \ I–I_2^- \text{ bond enthalpy} \\ - \ \text{lattice enthalpy of } I_2 \\ - \ \text{lattice enthalpy of } NaI \end{array}$$

The fourth term is larger than the first term, since $I^-$ is smaller than $I_3^-$. This difference provides the driving force for the reaction to occur as written. If we substitute $Cs^+$ for $Na^+$, the two middle terms will remain constant. Now, the fourth term is still larger than the first term, *but by a significantly smaller amount*. The net result is that $\Delta H$ for the cesium system is positive. In the limit where the cation becomes infinitely large, the difference between the first and fourth terms becomes negligible.

**13.11 Write Lewis structures and predict shapes: (a) $ClO_2$.** The Lewis structure and the predicted shape of $ClO_2$ are shown below. The angular shape is a consequence of repulsions between the bonding and nonbonding electrons. With three nonbonding electrons, the O–Cl–O angle in $ClO_2$ is 118°. With four nonbonding electrons, as in $ClO_2^-$, the repulsions are greater and the O–Cl–O angle is only 111°.

$$:\ddot{O} — \dot{C}l — \ddot{O}: \qquad\qquad \begin{array}{c} Cl \\ {}_O{}^{\diagup} \quad {}^{\diagdown}{}_O \end{array}$$

**(b) $I_2O_5$.** The Lewis structure and the predicted shape of $I_2O_5$ are shown below. The central O atom has two bonding pairs of electrons and two lone pairs, like $H_2O$, so it should be no surprise that the I–O–I bond angle is less than 180° (it is 139°). Each I atom is trigonal pyramidal, since it has three bonding pairs and one lone pair. If you had difficulty working this exercise, you should review VSEPR theory, covered in **Section 2.2**.

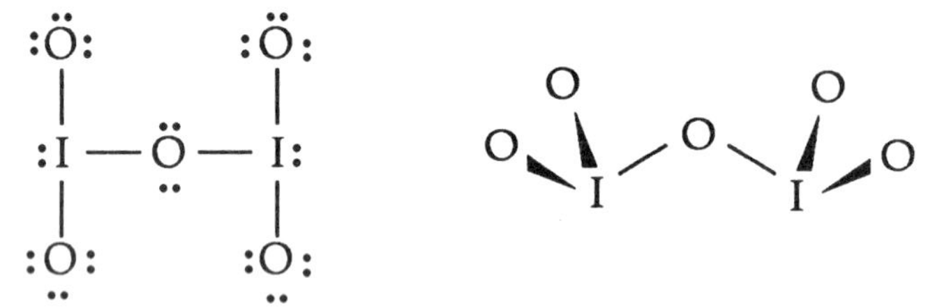

**13.12 The formulas and probable acidities of perbromic and periodic acid.** The formulas are $HBrO_4$ and $H_5IO_6$. The difference lies in iodine's ability to expand its coordination shell, a direct consequence of its large size. Recall from **Section 5.6** that the relative strength of an oxoacid can be estimated from the $q/p$ ratio ($q$ is the number of oxo groups and $p$ is the number of OH groups attached to the central atom). A high value of $q/p$ correlates with strong acidity while a low value correlates with weak acidity. For $HBrO_4$, $q/p$ = 3/1, so it is a strong acid. For $H_5IO_6$, $q/p$ = 1/5, so it is a weak acid. Periodic acid is thermodynamically more stable with respect to reduction than perbromic acid. Bromine, like its Period 4 neighbors arsenic and selenium, is more oxidizing in its highest oxidation state than the members of the group immediately above and below.

**13.13 Which reacts more quickly in dilute aqueous solution, perchloric acid or periodic acid?** Periodic acid is by far the quicker oxidant. Recall that it exists in two forms in aqueous solution, $H_5IO_6$ (the predominant form) and $HIO_4$. Even though the concentration of $HIO_4$ is low, this four-coordinate species can form a complex with a potential reducing agent, providing for an efficient and rapid mechanism by which the redox reaction can occur (see **Section 13.8**).

**13.14 Which oxo anions of chlorine undergo disproportionation in acid?**
The Frost diagram for chlorine in acid solution is shown below. It is worth

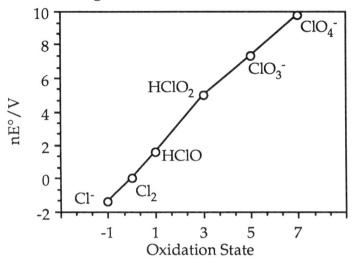

your while to construct this diagram yourself from the potential data in **Appendix 4**. Then, if you connect the points for $Cl^-$ and $ClO_4^-$ with a straight line, you will see that the inter- mediate oxidation state species $HClO$, $HClO_2$, and $ClO_3^-$ lie above that line. Therefore, they are unstable with respect to disproportionation. As discussed in **Section**
**13.7**, the redox reactions of halogen oxo anions become progressively faster as the oxidation number of the halogen *decreases*. Therefore, the rates of disproportionation are probably $HClO > HClO_2 > ClO_3^-$. Note that $ClO_4^-$ cannot undergo disproportionation, since there are no species with a higher oxidation number.

**13.15 Which of the following compounds present an explosion hazard:**
**(a) $NH_4ClO_4$?** The key to answering this question is to decide if the perchlorate ion, which is a very strong oxidant, is present along with a species that can be oxidized. If so, the compound *does* represent an explosion hazard. Ammonium perchlorate is a dangerous compound, since the N atom of the $NH_4^+$ ion is in its lowest oxidation state and can be oxidized. The explosion hazard of $NH_4ClO_4$ was mentioned in **Section 13.6**.

**(b) $Mg(ClO_4)_2$?** Since Mg(II) cannot be oxidized to a higher oxidation state, magnesium perchlorate is a stable compound and is not an explosion hazard.

(c) **NaClO$_4$.** The same answer applies here as above for magnesium perchlorate. Sodium has only one common oxidation state. Even the strongest oxidants, such as F$_2$, FOOF, and ClF$_3$, cannot oxidize Na(I) to Na(II).

(d) **[Fe(H$_2$O)$_6$][ClO$_4$]$_2$.** Although the H$_2$O ligands cannot be oxidized, the metal ion can. This compound presents an explosion hazard, since Fe(II) can be oxidized to Fe(III) by a strong oxidant such as perchlorate ion.

**13.16 Are the following likely to have simple ionic structures, layered structures, or metal-metal bonded structures: (a) FeF$_2$?** As discussed in **Section 13.9**, dihalides and trihalides of Period 3 $d$-block metals are ionic. The difluorides, such as FeF$_2$, typically exhibit the rutile structure, in which the metal ions are six-coordinate (see **Table 13.8**).

**(b) NiCl$_2$?** Unlike the difluorides of Period 3 $d$-block metals, the dichlorides, dibromides, and diiodides form lattices with less ionic and more covalent character. Thus, a layered structure is likely for NiCl$_2$.

**(c) ScCl$_2$?** Scandium(II) chloride is distinctly different from the other Period 3 $d$-block dichlorides. The only oxidation state exhibited by Sc in *ionic* compounds is +3, so ScCl$_2$ is really a subhalide. As such, it is likely to exhibit metal-metal bonding, as shown in **Figure 13.12** for Sc$_7$Cl$_{10}$.

**(d) WCl$_2$?** Many of the halides of Period 4 and Period 5 metals on the left of the $d$-block are metal-metal bonded compounds. Examples discussed in **Section 13.9** include compounds with empirical formulas ZrI$_2$ and ReCl$_3$. Therefore, by analogy it is likely that WCl$_2$ also contains metal-metal bonds.

**13.17 Give balanced equations and conditions for the synthesis of: (a) Iron(II) chloride tetrahydrate.** We can oxidize iron metal using an aqueous solution of Cl$_2$, as shown in the following equation:

$$2\text{Fe }(s) + \text{Cl}_2\,(aq) \rightarrow 2\text{Fe}^{2+}\,(aq) + 2\text{Cl}^-\,(aq)$$

Evaporation of the solution will cause crystallization of the tetrahydrate, FeCl$_2$·4H$_2$O. In planning this synthesis, we must be careful to use a limiting supply of chlorine to prevent the oxidation of Fe(II) to Fe(III). You should compare the $Fe^{3+}/Fe^{2+}$ and $Cl_2/Cl^-$ reduction potentials, found in **Appendix 4**, to confirm that this can occur.

**(b) Anhydrous $FeCl_2$.** We cannot treat iron metal with a stream of chlorine gas, because that would result in the formation of $FeCl_3$. Instead, we can prepare $FeCl_2 \cdot 4H_2O$, as above, and dehydrate it using thionyl chloride:

$$FeCl_2 \cdot 4H_2O \ (s) \ + \ 4SOCl_2 \ (l) \ \rightarrow \ FeCl_2 \ (s) \ + \ 4SO_2 \ (g) \ + \ 8HCl \ (g)$$

This type of reaction is usually performed using thionyl chloride as the solvent. The solid hydrated compound is treated with a large excess of $SOCl_2$ and heated to reflux. After a number of hours, the anhydrous compound is filtered and dried under an inert atmosphere.

**(c) $PtF_6$.** As discussed in **Section 13.10**, the direct combination of Pt and $F_2$ at a sustained high temperature cannot be used, because the product, $PtF_6$, is not stable at high temperatures. The diagram shown in **Figure 13.15** is one way around the problem. The coil of Pt metal is heated by passing an electrical current through it. It reacts with $F_2$ to form $PtF_6$. Since the heating is localized, the volatile product condenses on the cold walls of the reaction vessel. After the reaction is over, the reaction vessel is warmed to room temperature and the $PtF_6$ is purified by trap-to-trap distillation on a high-vacuum line (see **Box 9.1**).

**(d) ZrBr.** This subhalide of zirconium *cannot* be prepared by the direct combination of the elements, because that reaction only produces $ZrBr_4$:

$$Zr \ (s) \ + \ 2Br_2 \ (l) \ \rightarrow \ ZrBr_4 \ (s)$$

Instead, we must combine $ZrBr_4$ with Zr metal at high temperature in an inert vessel (it turns out that tantalum is the container of choice for such reactions, although this was not obvious to the first chemists that tried it):

$$ZrBr_4 \ (g) \ + \ 3Zr \ (s) \ \xrightarrow{\ 600-800 \ ^\circ C\ } \ 4ZrBr \ (s)$$

**13.18  Why is He rare in the earth's atmosphere?**  All of the original He and $H_2$ that was present in the earth's atmosphere when the earth originally formed has been lost.  Earth's gravitational field is not strong enough to hold these light gases, and they eventually diffuse away into space.  The small amount of helium that is present in today's atmosphere is the product of ongoing radioactive decay.

**13.19  Which of the noble gases would you choose as:  (a)  The lowest-temperature refrigerant?**  The noble gas with the lowest boiling point would best serve as the lowest-temperature refrigerant.  Helium, with a boiling point of 4.2 K, is the refrigerant of choice for very low temperature applications, such as cooling superconducting magnets in modern NMR spectrometers.  The boiling points of the noble gases are directly related to their size, or more correctly, to their number of electrons.  The larger atoms Kr and Xe are more polarizable than He or Ne, so Kr and Xe experience larger van der Waals attractions and have higher boiling points.

**(b)  An electric discharge light source requiring a safe gas with the lowest ionization potential?**  Of the "safe" noble gases, i.e., He–Xe, the largest one, Xe, has the lowest ionization potential (see **Table 13.1**).  Radon is even larger than Xe and has a lower ionization potential, but since it is radioactive it is not safe.

**(c)  The least expensive inert atmosphere?**  Geographic location might play a role in your answer to this question.  Since Ar is so plentiful in the atmosphere relative to the other noble gases, it is generally cheaper than helium, which is rare in the atmosphere.  However, a great deal of helium is collected as a byproduct of natural gas production.  In places where large amounts of natural gas are produced, such as the United States, helium may be marginally less expensive than argon.

**13.20  Give balanced equations and conditions for the synthesis of:  (a)  $XeF_2$.**  This compound can be prepared in two different ways.  First, a mixture of Xe and $F_2$ that contains an excess of Xe is heated to 400 °C.  The excess Xe prevents the formation of $XeF_4$ and $XeF_6$.  The second way is to photolyze a mixture of Xe and $F_2$ in a glass reaction vessel.  For either method of synthesis, the balanced equation is:

$$Xe\ (g)\ +\ F_2\ (g)\ \rightarrow\ XeF_2\ (s)$$

At 400 °C the product is a gas, but at room temperature it is a solid.

**(b) XeF$_6$.** For the synthesis of this compound we want to also use a high temperature, but unlike the synthesis of XeF$_2$, we want to have a *large* excess of F$_2$:

$$Xe\ (g)\ +\ 3F_2\ (g)\ \rightarrow\ XeF_6\ (s)$$

As with XeF$_2$, xenon hexafluoride is a solid at room temperature.

**(c) XeO$_3$.** This compound is endoergic, so it cannot be prepared directly from the elements. However, a sample of XeF$_6$ can be carefully hydrolyzed to form the product:

$$XeF_6\ (s)\ +\ 3H_2O\ (l)\ \rightarrow\ XeO_3\ (s)\ +\ 6HF\ (g)$$

If a large excess of water is used, an aqueous solution of XeO$_3$ is formed instead.

**13.21  What noble gas species is isostructural with: (a) ICl$_4^-$?** The Lewis structure of this anion has four bonding pairs and two lone pairs of electrons around the central iodine atom. Therefore, the anion has a square planar geometry. The noble gas compound that is isostructural is XeF$_4$, as shown below.

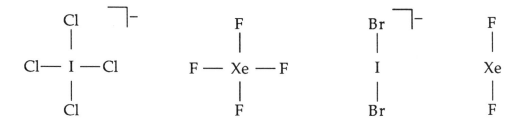

**(b) IBr$_2^-$?** The Lewis structure of this anion has two bonding pairs and three lone pairs of electrons around the central iodine atom. Therefore, the anion has a linear geometry. The noble gas compound that is isostructural is $XeF_2$, as shown above.

(c) $BrO_3^-$? The Lewis structure of this anion has three bonding pairs and one lone pair of electrons around the central bromine atom. Therefore, the anion has a trigonal pyramidal geometry. The noble gas compound that is isostructural is $XeO_3$, as shown below.

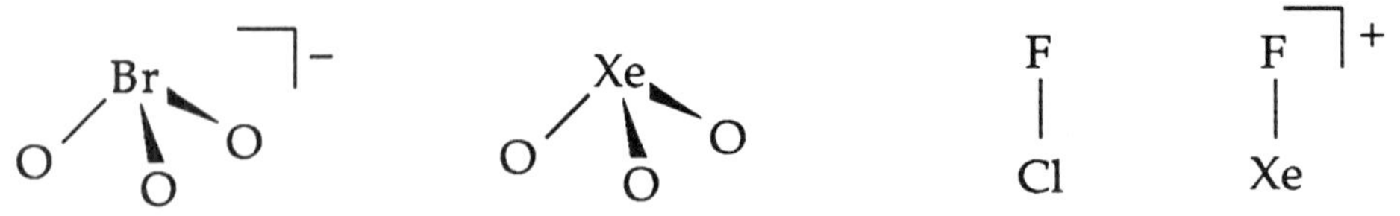

(d) ClF? This diatomic molecule does not have a *molecular* noble gas counterpart (i.e., all noble gas compounds have three or more atoms). However, it is isostructural with the cation $XeF^+$, as shown above.

**13.22 With what neutral molecules is ClO⁻ isoelectronic? Are these molecules Lewis acids?** This anion is isoelectronic with the diatomic halogen molecules ($F_2$, $Cl_2$, $Br_2$, and $I_2$) and with the diatomic interhalogen molecules (ClF, IF, ICl, IBr, etc.). These molecules are indeed Lewis acids, as discussed in **Sections 13.4 and 6.8**. Therefore, it is reasonable to assume that the chlorine end of ClO⁻ can act as an acceptor. If so, the disproportionation of ClO⁻ might proceed by attack of the oxygen end of ClO⁻ (the nucleophile) on the chlorine end of another ClO⁻ anion (the electrophile):

$$ClO^- + ClO^- \rightarrow [ClO\cdots ClO]^{2-} \rightarrow [Cl\cdots OClO]^{2-} \rightarrow Cl^- + ClO_2^-$$

# CHAPTER 14

# BONDING AND SPECTRA OF COMPLEXES

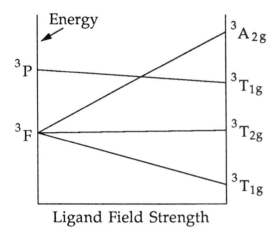

A correlation diagram between a free $d^2$ metal ion (left side) and the octahedral strong-field terms of a $d^2$ metal ion (right side). Only the spin triplet terms are shown. There are three spin-allowed transitions between the $^3T_{1g}$ ground state and higher energy spin triplet states, $^3T_{2g} \leftarrow {}^3T_{1g}$, $^3T_{1g} \leftarrow {}^3T_{1g}$, and $^3A_{2g} \leftarrow {}^3T_{1g}$.

## SOLUTIONS TO IN-CHAPTER EXERCISES

**14.1   What terms arise from an $s^1d^1$ configuration?** An atom with an $s^1d^1$ configuration will have $L = 0 + 2 = 2$ (an electron in an $s$ orbital has $l = 0$, while an electron in a $d$ orbital has $l = 2$). This value of $L$ corresponds to a D term. Since the electrons may be paired ($S = 0$, multiplicity $2S + 1 = 1$) or parallel ($S = 1$, multiplicity $2S + 1 = 3$), both $^1D$ and $^3D$ terms are possible.

**14.2   Identifying ground terms. (a) $2p^2$.** Two electrons in a $p$ subshell can occupy separate $p$ orbitals and have parallel spins ($S = 1$), so the maximum multiplicity $2S + 1 = 3$. The maximum value of $M_L$ is $1 + 0 = 1$, which corresponds to a P term (according to the Pauli principle, $m_l$ cannot be +1 for both electrons if the spins are parallel). Thus, the ground term is $^3P$ (called "a triplet P" term).

**(b) $3d^9$.** The largest value of $M_S$ for nine electrons in a $d$ subshell is 1/2 (eight electrons are paired in four of the five orbitals, while the ninth electron has $m_s = \pm 1/2$). Thus, $S = 1/2$ and the multiplicity $2S + 1 = 2$. Notice that the largest value of $M_S$ is the same for one electron in a $d$ subshell. Similarly, the largest value of $M_L$ is 2, which results from the following nine values of $m_l$: +2, +2, +1, +1, 0, 0, −1, −1, −2. Notice that $M_L = 2$ also corresponds to one electron in a $d$ subshell. Thus, the ground term for a $d^9$ or a $d^1$ configuration is $^2D$ (called "a doublet D" term).

**14.3   What terms in a $d^2$ complex of $O_h$ symmetry correlate with the $^3F$ and $^1D$ terms of the free ion?** According to **Table 14.3**, an F term arising from a $d^n$ configuration correlates with $T_{1g} + T_{2g} + A_{2g}$ terms in an $O_h$ complex. The multiplicity is unchanged by the correlation, so the terms in $O_h$ symmetry are $^3T_{1g}$, $^3T_{2g}$, and $^3A_{2g}$. Similarly, a D term arising from a $d^n$ configuration correlates with $T_{2g} + E_g$ terms in an $O_h$ complex. The $O_h$ terms retain the singlet character of the $^1D$ free ion term, and so are $^1T_{2g}$ and $^1E_g$.

**14.4   Predict the wavenumbers of the first two spin-allowed bands in the spectrum of $[Cr(H_2O)_6]^{3+}$.** The two transitions in question are $^4T_2 \leftarrow {}^4A$ and $^4T_1 \leftarrow {}^4A$ (the first one has a lower energy and hence a lower wavenumber). Since $\Delta_o = 17,600$ cm$^{-1}$ and $B = 918$ cm$^{-1}$, we must find the point $17,600/918 = 19.2$ on the x-axis in **Figure 14.5**. Then, the ratios $E/B$ for the two bands are the y values of the points on the $^4T_2$ and $^4T_1$ lines, 20 and 30, respectively, as shown below. Therefore, the two lowest-energy spin-allowed bands in the spectrum of $[Cr(H_2O)_6]^{3+}$ will be found at (918 cm$^{-1}$)(20) = 18,400 cm$^{-1}$ and (918 cm$^{-1}$)(30) = 27,500 cm$^{-1}$.

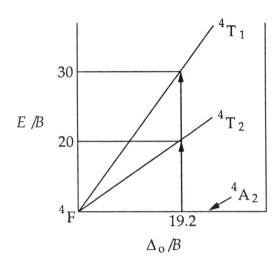

A simplified version of the Tanabe-Sugano diagram for the $d^3$ configuration of $[Cr(H_2O)_6]^{3+}$.

**14.5   Assign the bands in the spectrum of $[Cr(NCS)_6]^{3-}$.** This six-coordinate $d^3$ complex undoubtedly has $O_h$ symmetry, so the general features of its spectrum will resemble the spectrum of $[Cr(NH_3)_6]^{3+}$, shown in **Figure 14.1**. The very low-intensity of the band at 16,000 $cm^{-1}$ is a clue that it is a spin-forbidden transition, probably $^2E_g \leftarrow {}^4A_{2g}$. Spin-allowed but Laporte-forbidden bands typically have $\varepsilon \sim 100$ $M^{-1}$ $cm^{-1}$, so it is likely that the bands at 17,700 $cm^{-1}$ and 23,800 $cm^{-1}$ are of this type (they correspond to the $^4T_{1g} \leftarrow {}^4A_{2g}$ and $^4T_{2g} \leftarrow {}^4A_{2g}$ transitions, respectively). The band at 32,400 $cm^{-1}$ is probably a charge transfer band, since its intensity is too high to be a ligand field ($d$-$d$) band. Since we are provided with the hint that the $NCS^-$ ligands have low-lying $\pi^*$ orbitals, it is reasonable to conclude that this band corresponds to a MLCT transition. Notice that the two spin-allowed ligand field transitions of $[Cr(NCS)_6]^{3-}$ are at lower energy than those of $[Cr(NH_3)_6]^{3+}$, showing that $NCS^-$ induces a smaller $\Delta_0$ on $Cr^{3+}$ than does $NH_3$. Also notice that $[Cr(NH_3)_6]^{3+}$ lacks an intense MLCT band at ~30,000-40,000 $cm^{-1}$, showing that $NH_3$ does not have low-lying empty orbitals.

**14.6   What evidence from absorption and CD spectra show that $[Co(edta)]^-$ has lower symmetry than $[Co(en)_3]^{3+}$?** The evidence is the greater number of CD bands in the spectrum of $[Co(edta)]^-$ than in the spectrum of $[Co(en)_3]^{3+}$. The CD spectrum of $[Co(en)_3]^{3+}$, shown in **Figure 14.11**, exhibits three bands, near 21,000 $cm^{-1}$, 24,000 $cm^{-1}$, and 26,000 $cm^{-1}$. The spectrum of $[Co(edta)]^-$, shown in **Figure 14.13**, exhibits five bands, near 17,000 cm$^{-1}$, 19,000 cm$^{-1}$, 24,000 cm$^{-1}$, 26,000 cm$^{-1}$, and 28,000 cm$^{-1}$. The greater number of bands for [Co(edta)]$^-$ results from a lifting of degeneracies, which implies that this complex has a lower symmetry than [Co(en)$_3$]$^{3+}$.

**14.7   Electrical properties of magnetite.** Magnetite contains both $Fe^{2+}$ and $Fe^{3+}$ in octahedral sites, so type III metallic behavior might be expected. This would give rise to good electrical conductivity, which is observed for this mixed valence compound.

| SOLUTIONS TO END-OF-CHAPTER EXERCISES |
| --- |

**14.1   Russell-Saunders term symbols.   (a)** $L = 0$, $S = 5/2$. You should remember that a term, denoted by a capital letter, is related to $L$ in the same way that an orbital, denoted by a lower case letter, is related to $l$:

| If $l$ = | 0 | 1 | 2 | 3 | 4 | 5 | 6 |
| --- | --- | --- | --- | --- | --- | --- | --- |
| orbital = | $s$ | $p$ | $d$ | $f$ | $g$ | $h$ | $i$ |

| If $L$ = | 0 | 1 | 2 | 3 | 4 | 5 | 6 |
| --- | --- | --- | --- | --- | --- | --- | --- |
| term = | S | P | D | F | G | H | I |

The multiplicity of the term, which is always given as a left superscript, can always be determined by using the formula multiplicity = $2S + 1$:

| If $S$ = | 0 | 1/2 | 1 | 3/2 | 2 | 5/2 |
| --- | --- | --- | --- | --- | --- | --- |
| multiplicity $2S + 1$ = | 1 | 2 | 3 | 4 | 5 | 6 |

The terms and multiplicities listed above are not a complete list to cover all possibilities for all atoms and ions, but they will cover all possible $d^n$ configurations. As far as the situation $L = 0$, $S = 5/2$ is concerned, the term symbol is $^6S$. In addition to answering this question, you should try to decide which $d^n$ configurations can give rise to the term. In this case, the only $d^n$ configuration that can give rise to an $^6S$ term is $d^5$ (e.g., a gas-phase $Mn^{2+}$ or $Fe^{3+}$ ion).

**(b)** $L = 3$, $S = 3/2$. According to the relations shown above, this set of angular momentum quantum numbers is described by the term symbol $^4F$. The diagram below, which is another way to depict a microstate, shows how the situation $L = 3$, $S = 3/2$ can arise from a $d^3$ configuration (e.g., a gas-phase $Cr^{3+}$ ion). It can also arise from $d^5$ and $d^7$ configurations.

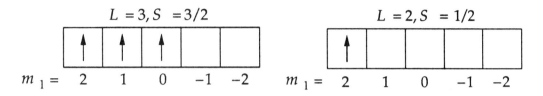

**(c)** $L = 2$, $S = 1/2$. This set of quantum numbers is described by the term symbol $^2D$. The microstate diagram above shows how the situation $L = 2$, $S = 1/2$ can arise from a $d^1$ configuration (e.g., a gas-phase $Ti^{3+}$ ion). It can also arise from $d^3$, $d^5$, $d^7$, and $d^9$ configurations.

**(d)** $L = 1$, $S = 1$. This set of quantum numbers is described by the term symbol $^3P$. It can arise from $d^2$, $d^4$, $d^6$, and $d^8$ configurations.

**14.2   Identify the ground term. (a) $^3F$, $^3P$, $^1P$, $^1G$.** By definition, the ground term has the lowest energy of all of the terms. Recall Hund's two rules, discussed in **Section 14.2**: (1) The term with the greatest multiplicity lies lowest in energy; (2) For a given multiplicity, the greater the value of $L$ of a term, the lower the energy. Therefore, the ground term in this case will be a triplet, not a singlet (rule 1). Of the two triplet terms, $^3F$ lies lower in energy than $^3P$: $L = 3$ for $^3F$, $L = 1$ for $^3P$ (rule 2). Therefore, the ground term is $^3F$.

**(b) $^5D$, $^3H$, $^3P$, $^1I$.** The ground term is $^5D$ because this term has a higher multiplicity than the other terms.

**(c) $^6S$, $^4G$, $^4P$, $^2I$.** The ground term is $^6S$ because this term has a higher multiplicity than the other terms.

**14.3   Give the Russell-Saunders terms for the following configurations and identify the ground term: (a) $4s^1$.** We can approach this Exercise in the way described in **Section 14.2** for the $d^2$ configuration (see **Table 14.1**). We write down all possible microstates for the $s^1$ configuration, then write down the $M_L$ and $M_S$ values for each microstate, then infer the values of $L$ and $S$ to which the microstates belong. In this case, the procedure is not lengthy, since there are only two possible microstates, $(0^+)$ and $(0^-)$. The only possible values of $M_L$ and $M_S$ are 0 and 1/2, respectively. If $M_L$ can only be 0, then $L$ must be 0, which gives an S term (remember that $L$ can take on all values $M_L, M_L - 1, ..., 0, ...-M_L$). Similarly, if $M_S$ can only be 1/2 or –1/2, then $S$ must be 1/2, which gives a multiplicity $2S + 1 = 2$. Therefore, the one and only term that arises from a $4s^1$ configuration is $^2$S.

   **(b) $3p^2$.** This case is more complicated because there are 15 possible microstates:

|  | $M_S = -1$ | $M_S = 0$ | $M_S = 1$ |
|---|---|---|---|
| $M_L = 2$ |  | $(1^+, 1^-)$ |  |
| $M_L = 1$ | $(1^-, 0^-)$ | $(1^+, 0^-), (1^-, 0^+)$ | $(1^+, 0^+)$ |
| $M_L = 0$ | $(1^-, -1^-)$ | $(1^+, -1^-), (0^+, 0^-), (-1^+, 1^-)$ | $(1^+, -1^+)$ |
| $M_L = -1$ | $(-1^-, 0^-)$ | $(-1^+, 0^-), (-1^-, 0^+)$ | $(-1^+, 0^+)$ |
| $M_L = -2$ |  | $(-1^+, -1^-)$ |  |

A term that contains a microstate with $M_L = 2$ must be a D term ($L = 2$). This term can only be a singlet, since if both electrons have $m_l = 1$ they must be spin-paired. Therefore, one of the terms of the $3p^2$ configuration is $^1$D, which contains five microstates. To account for these, we can cross out $(1^+, 1^-), (-1^+, -1^-)$, and one microstate from each of the other three rows under $M_S = 0$. That leaves ten microstates to be accounted for. The maximum value of $M_L$ of the remaining microstates is 1, so we next consider a P term ($L = 1$). Since $M_S$ can be –1, 0, or 1, this term will be $^3$P, which contains nine of the ten remaining microstates. The last remaining microstate is one of the original three for which $M_L = 0$ and $M_S = 0$, which is the one and only microstate that belongs to a $^1$S term ($L = 0$, $S = 0$). Of the three terms $^1$D, $^3$P, and $^1$S that arise from the $3p^2$ configuration, the ground term is $^3P$, since it has a higher multiplicity than the other two terms (see Hund's rule #1).

**14.4   Calculate the values of B and C for V$^{3+}$.** The diagram below shows the relative energies of the $^3F$, $^1D$, and $^3P$ terms:

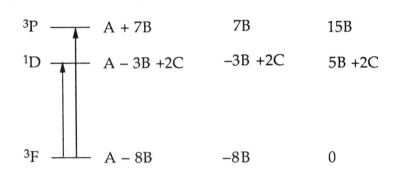

$$
\begin{array}{cccc}
^3P & \underline{\quad\quad} & A + 7B & 7B & 15B \\
\\
^1D & \underline{\quad\quad} & A - 3B + 2C & -3B + 2C & 5B + 2C \\
\\
\\
\\
^3F & \underline{\quad\quad} & A - 8B & -8B & 0
\end{array}
$$

Relative Energies

From the diagram it can be seen that the 10,642 cm$^{-1}$ energy gap between the $^3F$ and $^1D$ terms is 5B + 2C, while the 12,920 cm$^{-1}$ energy gap between the $^3F$ and $^3P$ terms is 15B.  From the two simultaneous equations

$$5B + 2C = 10{,}642 \text{ cm}^{-1} \quad \text{and} \quad 15B = 12{,}920 \text{ cm}^{-1}$$

we can determine that B = (12,920 cm$^{-1}$)/(15) = 861.33 cm$^{-1}$ and C = 3167.7 cm$^{-1}$.

**14.5   $d^n$ configurations and ground terms.   (a)   Low-spin [Rh(NH$_3$)$_6$]$^{3+}$.** There are a number of ways to determine the integer n for a given $d$-block metal.  One straightforward procedure is as follows.  Count the number of elements from the left side of the periodic table to the metal in question. This will be the number of $d$ electrons *for a metal atom in a complex* (note that an isolated gas-phase metal atom may have an $s^m d^n$ configuration, but the same neutral metal atom *in a complex* will have a $d^{m+n}$ configuration). Then subtract the positive charge on the metal ion from this number, leaving the integer n.  For example, Rh is the ninth element in period 5, so Rh$^0$ in a complex has a $d^9$ configuration, Rh$^+$ has a $d^8$ configuration, and so on.    Using this procedure, the $Rh^{3+}$ ion in the octahedral complex $[Rh(NH_3)_6]^{3+}$ has a $d^6$ configuration ($9 - 3 = 6$). According to the $d^6$ Tanabe-Sugano diagram (see **Appendix 7**), the ground term for a low-spin $t_{2g}^6$ metal ion is $^1A_{1g}$.

(b)  $[Ti(H_2O)_6]^{3+}$.  Titanium is the fourth element in period 4, so $Ti^0$ in a complex has a $d^4$ configuration.  Therefore, the $Ti^{3+}$ ion in the octahedral $[Ti(H_2O)_6]^{3+}$ ion has a $d^1$ configuration ($4 - 3 = 1$).  A correlation diagram for $d^1$ metal ions is shown in **Figure 14.2** (**Appendix 7** does not include the $d^1$ Tanabe-Sugano diagram).  According to this diagram, the ground term for a $t_{2g}^1$ metal ion is $^2T_{2g}$.

(c)  **High-spin $[Fe(H_2O)_6]^{3+}$.**  Iron is the eighth element in period 4, so $Fe^0$ in a complex (such as $Fe(CO)_5$) has a $d^8$ configuration.  Therefore, the $Fe^{3+}$ ion in the octahedral $[Ti(H_2O)_6]^{3+}$ ion has a $d^5$ configuration ($8 - 3 = 5$). According to the $d^5$ Tanabe-Sugano diagram, the ground term for a high-spin $t_{2g}^3 e_g^2$ metal ion is $^6A_{1g}$.

**14.6**    Estimate $\Delta_o$ and $B$ for: (a) $[Ni(H_2O)_6]^{2+}$.  According to the $d^8$ Tanabe-Sugano diagram (**Appendix 7**), the absorptions at 8,500 $cm^{-1}$, 13,800 $cm^{-1}$, and 25,300 $cm^{-1}$ correspond to the following spin-allowed transitions, respectively: $^3T_{2g} \leftarrow {^3A_{2g}}$, $^3T_{1g} \leftarrow {^3A_{2g}}$, and $^3T_{1g} \leftarrow {^3A_{2g}}$.  The ratios $13,800/8,500 = 1.6$ and $25,300/8500 = 3.0$ can be used to estimate $\Delta_o/B \approx 11$.  Using this value of $\Delta_o/B$  and the fact that $E/B = \Delta_o/B$ for the lowest-energy transition, $\Delta_o = 8,500$ $cm^{-1}$ and $B \approx 770$ $cm^{-1}$.  Note that $B$ for a gas-phase $Ni^{2+}$ ion is 1080 $cm^{-1}$ (see **Table 14.2**).  The fact that $B$ for the complex is only ~70% of the free ion value is an example of the nephalauxetic effect (**Section 14.3**).

(b)  $[Ni(NH_3)_6]^{2+}$.  The absorptions for this complex are at 10,750 $cm^{-1}$, 17,500 $cm^{-1}$, and 28,200 $cm^{-1}$.  The ratios in this case are $17,500/10,750 = 1.6$ and $28,200/10,750 = 2.6$, and lead to $\Delta_o/B \approx 15$.  Thus, $\Delta_o = 10,750$ $cm^{-1}$ and $B \approx 720$ $cm^{-1}$.  It is sensible that $B$  for  $[Ni(NH_3)_6]^{2+}$  is smaller than $B$ for $[Ni(H_2O)_6]^{2+}$, since $NH_3$ is higher in the nephalauxetic series than is $H_2O$.

**14.7**    **Ground term and lowest energy transition for a paramagnetic Fe(II) complex.**  A $d^6$ octahedral Fe(II) complex can be either high-spin ($t_{2g}^4 e_g^2$, $S =$

2) or low-spin ($t_{2g}^6$, $S = 0$). If the complex has a large paramagentic suscepti-bility, it must be high-spin, since a low-spin complex would be diamagnetic. According to the $d^6$ Tanabe-Sugano diagram (**Appendix 7**), the ground term for the high-spin case (i.e., to the left of the discontinuity) is $^5T_{2g}$. The only other quintet term is $^5E_g$, so the only spin-allowed transition is $^5E_g \leftarrow ^5T_{2g}$.

**14.8   The spectrum of [Co(NH$_3$)$_6$]$^{3+}$.** If this $d^6$ complex were high-spin, the only spin-allowed transition possible would be $^5E_g \leftarrow ^5T_{2g}$ (refer once again to the $d^6$ Tanabe-Sugano diagram). On the other hand, if it were low-spin, several spin-allowed transitions are possible, including $^1T_{1g} \leftarrow ^1A_{1g}$, $^1T_{2g} \leftarrow ^1A_{1g}$, $^1E_g \leftarrow ^1A_{1g}$, etc. The presence of *two* moderate-intensity bands in the visible/near-UV spectrum of [Co(NH$_3$)$_6$]$^{3+}$ suggests that it is low-spin. The first two transitions listed above correspond to these two bands. The very weak band in the red corresponds to a spin-forbidden transition such as $^3T_{2g} \leftarrow ^1A_{1g}$.

**14.9   Explain why [FeF$_6$]$^{3-}$ is colorless whereas [CoF$_6$]$^{3-}$ is colored.** The $d^5$ Fe$^{3+}$ ion in the octahedral hexafluoro complex must be high-spin. According to the $d^5$ Tanabe-Sugano diagram (**Appendix 7**), a high-spin complex has no higher energy terms of the same multiplicity as the $^6A_{1g}$ ground term. Therefore, since no spin-allowed transitions are possible, the complex is expected to be colorless (i.e., only very weak spin-forbidden transitions are possible). If this Fe(III) complex were low-spin, spin-allowed transitions such as $^2T_{1g} \leftarrow ^2T_{2g}$, $^2A_{2g} \leftarrow ^2T_{2g}$, etc. would render the complex colored. The $d^6$ Co$^{3+}$ ion in [CoF$_6$]$^{3-}$ is also high-spin, but in this case a single spin-allowed transition, $^5E_g \leftarrow ^5T_{2g}$, makes the complex colored and gives it a one-band spectrum.

**14.10   Explain why the nephalauxetic effect for CN$^-$ is larger than for NH$_3$.** These two ligands are quite different with respect to the types of bonds they form with metal ions. Ammonia and cyanide ion are both σ-bases, but cyanide is also a π-acid. This difference means that NH$_3$ can form molecular orbitals only with the metal $e_g$ orbitals, while CN$^-$ can form molecular orbitals with the metal $e_g$ and $t_{2g}$ orbitals. The formation of molecular orbitals is way that ligands "expand the clouds" of the metal $d$ orbitals.

**14.11  The origins of transitions for a complex of Co(III) with ammine and chloro ligands.** Lets start with the intense band at relatively high energy with $\varepsilon_{max} = 2 \times 10^4$ M$^{-1}$ cm$^{-1}$. This is undoubtedly a spin-allowed charge-transfer transition, since it is too intense to be a ligand field ($d$-$d$) transition. Furthermore, it is probably a LMCT transition, not a MLCT transition, since the ligands do not have the empty orbitals necessary for a MLCT transition. The two bands with $\varepsilon_{max} = 60$ and $80$ M$^{-1}$ cm$^{-1}$ are probably spin-allowed ligand field transitions. Even though the complex is not strictly octahedral, the ligand field bands are still not very intense. The *very* weak peak with $\varepsilon_{max} = 2$ M$^{-1}$ cm$^{-1}$ is most likely a spin-forbidden ligand field transition.

**14.12  Describe the transitions of Fe$^{3+}$ impurities in bottle glass.** The Fe$^{3+}$ ions in question are $d^5$ metal ions. If they were low-spin, several spin-allowed ligand field transitions would give the glass a color even when viewed through the wall of the bottle (see the Tanabe-Sugano diagram for $d^5$ metal ions in **Appendix 7**). Therefore, the Fe$^{3+}$ ions are high-spin, and as such have no spin-allowed transitions (the ground state of an octahedral high-spin $d^5$ metal ion is $^6A_{1g}$, and there are no sextet excited states). The faint green color, which is only observed when looking through a *long* pathlength of bottle glass, is due to spin-forbidden ligand field transitions.

**14.13  The origins of transitions for [Cr(H$_2$O)$_6$]$^{3+}$ and CrO$_4{}^{2-}$.** The pale violet color of the Cr$^{3+}$ ions in [Cr(H$_2$O)$_6$]$^{3+}$ is due to spin-allowed but Laporte-forbidden ligand field transitions. The relatively low molar absorption coefficient, $\varepsilon$, which is a manifestation of the Laporte-forbidden nature of the transitions, is the reason that the color is "pale." The oxidation state of chromium in dichromate dianion is Cr(VI), which is $d^0$. Therefore, no ligand field transitions are possible. The intense yellow color is due to LMCT transitions (i.e., electron transfer from the oxide ion ligands to the Cr(VI) metal center). Charge transfer transitions are intense because they are both spin-allowed and Laporte-allowed.

**14.14  The orbitals of [Co(NH$_3$)$_5$Cl]$^{2+}$.** The $d_{z^2}$ orbital in this complex is left unchanged by each of the symmetry operations of the $C_{4v}$ point group. It therefore has A$_1$ symmetry. The Cl atom lone pairs of electrons can form $\pi$

molecular orbitals with $d_{xz}$ and $d_{yz}$. These metal atomic orbitals are $\pi$-anti-bonding MOs in $[Co(NH_3)_5Cl]^{2+}$ (whereas they are nonbonding in $[Co(NH_3)_6]^{3+}$), and so they will be raised in energy relative to their position in $[Co(NH_3)_6]^{3+}$, in which they were degenerate with $d_{xy}$. Since $Cl^-$ ion is not as strong a $\sigma$ base as $NH_3$, the $d_{z^2}$ orbital in $[Co(NH_3)_5Cl]^{2+}$ will be at lower energy than in $[Co(NH_3)_6]^{3+}$, in which it was degenerate with $d_{x^2-y^2}$. A qualitative $d$-orbital splitting diagram for both complexes is shown below (L $= NH_3$).

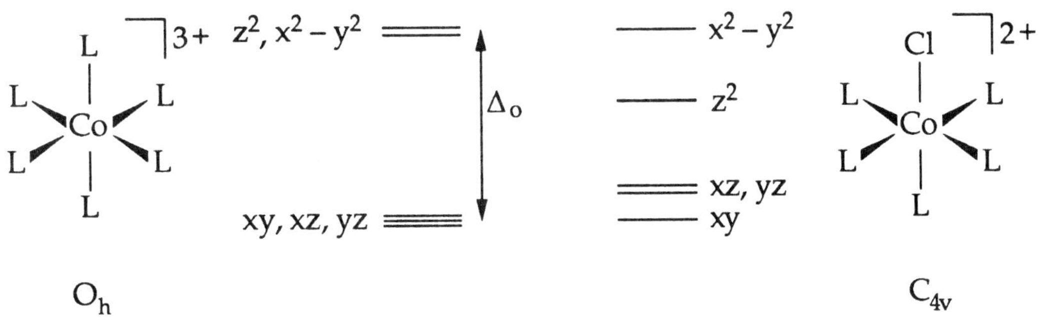

**14.15 Show that the purple color of $MnO_4^-$ cannot arise from a ligand field transition.** Reference to **Figure 7.16** shows that ligand field transitions can occur for $d$-block metal ions with one or more, but fewer than ten, electrons in the metal $e$ and $t_2$ orbitals. However, the oxidation state of manganese in permanganate anion is Mn(VII), which is $d^0$. Therefore, no ligand field transitions are possible. The metal $e$ orbitals are the LUMOs, and can act as acceptor orbitals for LMCT transitions. These fully allowed transitions give permanganate its characteristic *intense* purple color.

**14.16 Explain how to estimate $\Delta_T$ for $MnO_4^-$.** Referring again to **Figure 7.16**, you can see that two possible LMCT transitions are possible, $e \leftarrow a_1$ and $t_2 \leftarrow a_1$. These two transitions give rise to the two absorption bands observed at 18,500 cm$^{-1}$ and 32,200 cm$^{-1}$. The difference in energy between the two transitions, $E(t_2) - E(e) = 13,700$ cm$^{-1}$, is just equal to $\Delta_T$, as shown in the diagram below:

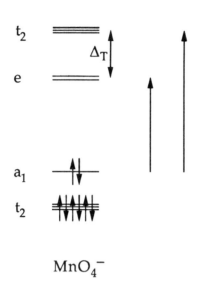

energies of the two
charge transfer
transitions

# CHAPTER 15

# REACTION MECHANISMS OF
# *d*-BLOCK COMPLEXES

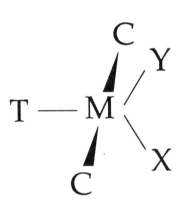

The proposed five-coordinate inter-mediate in substitution reactions of square-planar complexes. The incoming ligand Y adds to the four-coordinate complex $MC_2TX$ (X is the leaving group; T is *trans* to X; the C ligands are *cis* to X). The trigonal bipyramid that forms has T, X, and Y in the equatorial plane. The electronic properties of T dramatically affect the rate of the reaction (the *trans* effect), while the electronic properties of C do not.

---

## SOLUTIONS TO IN-CHAPTER EXERCISES

**15.1    Calculate the second order rate constant for the reaction of *trans*-[PtCl(CH₃)(PEt₃)₂] with NO₂⁻ in MeOH.**  We can make use of **Equation 4**:

$$\log k_2(NO_2^-) = Sn_{Pt}(NO_2^-) + C$$

where $S$ is the nucleophilic discrimination factor for this complex, $n_{Pt}(NO_2^-)$ is the nucleophilicity parameter of nitrite ion, and $C$ is the logarithm of the second order rate constant for the substitution of $Cl^-$ in this complex by MeOH. $S$ was determined to be 0.41 in the **Example**, and $n_{Pt}(NO_2^-)$ is given as 3.22. $C$ is a constant for a given complex and is $-0.61$ in this case. Therefore, $k_2(NO_2^-)$ can be determined as follows:

$$\log k_2(NO_2^-) = (0.41)(3.22) - 0.61 = 0.71$$

$$k_2(NO_2^-) = 10^{0.71} = 5.1 \; M^{-1} \; s^{-1}$$

**15.2   Propose efficient routes to *cis*- and *trans*-[PtCl$_2$(NH$_3$)(PPh$_3$)].** For the three ligands in question, $Cl^-$, $NH_3$, and $PPh_3$, the *trans* effect series is $NH_3 <$ $Cl^- < PPh_3$. This means that a ligand *trans* to $Cl^-$ will be substituted at a faster rate than a ligand *trans* to $NH_3$, and a ligand *trans* to $PPh_3$ will be substituted at a faster rate than a ligand *trans* to $Cl^-$. Since our starting material is $[PtCl_4]^{2-}$, two steps involving substitution *of* $Cl^-$ *by* $NH_3$ or $PPh_3$ must be used. If we first add $NH_3$ to $[PtCl_4]^{2-}$, we will produce $[PtCl_3(NH_3)]^-$. Now if we add $PPh_3$, one of the mutually *trans* $Cl^-$ ligands will be substituted faster than the $Cl^-$ ligand *trans* to $NH_3$, and the *cis* isomer will be the result:

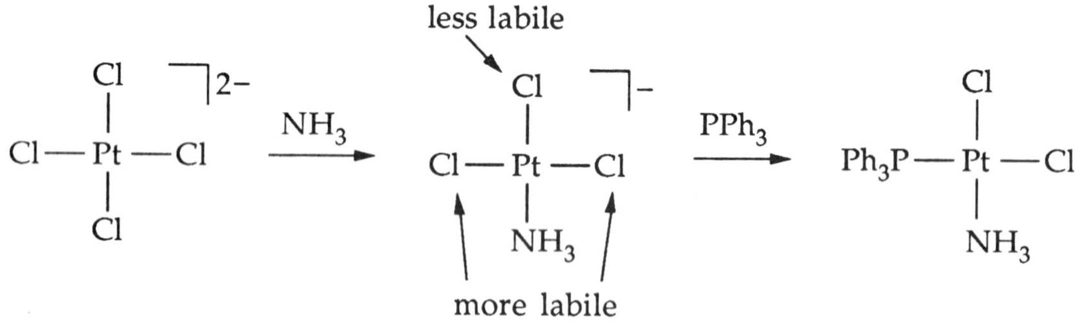

If we first add $PPh_3$ to $[PtCl_4]^{2-}$, we will produce $[PtCl_3(PPh_3)]^-$. Now if we add $NH_3$, the $Cl^-$ ligand *trans* to $PPh_3$ will be substituted faster than one of the mutuallyt rans $Cl^-$ ligands, and the *trans* isomer will be the result:

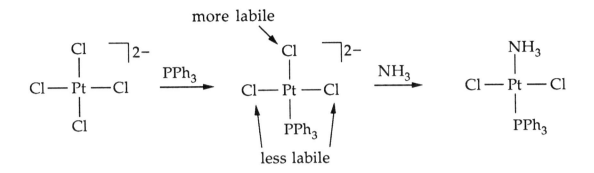

**15.3 Calculate *k* for the reaction of $[V(H_2O)_6]^+$ with Cl⁻.** As discussed in **Section 15.5**, the Eigen-Wilkins mechanism for substitution in octahedral complexes is:

$$[V(H_2O)_6]^{2+} + Cl^- \underset{}{\overset{K_E}{\rightleftharpoons}} \{[V(H_2O)_6]^{2+}, Cl^-\}$$

$$\{[V(H_2O)_6]^{2+}, Cl^-\} \xrightarrow{k} [VCl(H_2O)_5]^+ + H_2O$$

The observed rate constant, $k_{obs}$, is given by:

$$k_{obs} = kK_E$$

and in the case of the substitution of $H_2O$ by Cl⁻ in $[V(H_2O)_6]^{2+}$ is $1.2 \times 10^2$ $M^{-1}$ $s^{-1}$. We will be able to calculate *k* if we can estimate a proper value for $K_E$. Inspection of **Table 15.6** shows that $K_E = 1$ $M^{-1}$ for the encounter complex formed by F⁻ or SCN⁻ and $[Ni(H_2O)_6]^{2+}$. This value can be used for the reaction in question since (i) the charge and size of Cl⁻ are similar to those of F⁻ and SCN⁻ and (ii) the charge and size of $[Ni(H_2O)_6]^{2+}$ are similar to those of $[V(H_2O)_6]^{2+}$. Therefore, $k = k_{obs}/K_E = (1.2 \times 10^2$ $M^{-1}$ $s^{-1})/(1$ $M^{-1}) = 1.2 \times 10^2$ $s^{-1}$.

**15.4 What is the expected *cis-trans* ratio for the $[Co(NH_3)_4A(OH_2)]^{2+}$?** The trigonal bipyramidal intermediate shown in the lower path of **Figure 15.6** is reproduced below (O = $NH_3$). The three possible positions of attack in the equatorial plane are between the A and $NH_3$ ligands (there are two such positions) and between two $NH_3$ ligands (there is one such position). If the entering group Y ($H_2O$) can randomly attack these three positions, twice as much *cis* product as *trans* product will be formed. The *cis-trans* ratio will be 2:1.

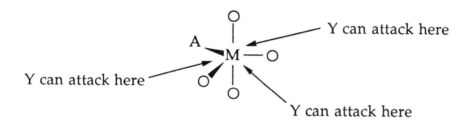

**15.5   What is $k$ if the overall reaction has $E° = 1.00$ V?**   If the reaction has $E° = 1.00$ V, the equlibrium constant $K$ will not be the same is in the **Example**. A reaction potential of 1.00 V will lead to a free energy change of $-(1.602 \times 10^{-19}$ J/eV$)(6.023 \times 10^{23}$ mol$^{-1}) = -9.65 \times 10^4$ J mol$^{-1}$. Knowing this, $K$ can be calculated as follows ($T = 273$ K):

$$K = \exp(-\Delta G/RT) = \exp((9.65 \times 10^4)/(8.31 \text{ J mol}^{-1} \text{ K}^{-1})(273 \text{ K}))$$

$$K = {} = e^{42.5} = 2.97 \times 10^{18}$$

Setting $f = 1$ gives

$$k = (9.0 \times 48 \times 2.97 \times 10^{18})^{1/2} \text{ M}^{-1} \text{ s}^{-1} = 3.6 \times 10^{10} \text{ M}^{-1} \text{ s}^{-1}$$

---

## SOLUTIONS TO END-OF-CHAPTER EXERCISES

**15.1   Classify the following as nucleophiles or electrophiles:  (a) $NH_3$.** Nucleophiles are intrinsically basic species (i.e., Lewis bases), while electrophiles are intrinsically acidic species (i.e., Lewis acids). A Lewis base may be either a good nucleophile or a poor nucleophile, but a base is a nucleophile. Correspondingly, a Lewis acid may be either a good electrophile or a poor electrophile, but it is rarely a nucleophile (a few ligands that are ambiphilic, like $SO_2$, display Lewis acidity *and* are nucleophiles). Within the framework of these definitions, $NH_3$ is a nucleophile, since it is a Lewis base.

   **(b) Cl⁻.** Chloride ion is a nucleophile, since it is a Lewis base (albeit a weak one).

   **(c) Ag⁺.** Silver(I) ion is an electrophile, since it is a Lewis acid.

   **(d) S²⁻.** Sulfide ion is a nucleophile, since it is a Lewis base.

   **(e) Al³⁺.** Aluminum(III) ion is an electrophile, since it is a Lewis acid.


**15.2   Is the reaction of $Ni(CO)_4$ with phosphines and phosphites *d* or *a*?** Since the rate of substitution is the same for a variety of different entering ligands L, the activated complex in each case must not include any significant bond making to the entering ligand and the reaction must be *d*. If the rate determining step included any Ni–L bond making, the rate of substitution would change as the electronic and steric properties of L were changed.


**15.3   Draw a reaction profile for a *D* reaction in which the rate determining step is the addition of the entering group Y to the intermediate.** The diagram below is such a reaction profile. The fact that the intermediate Y,M,X has a lower coordination number than either M–X or M–Y makes this a *D* reaction. The height of the free energy barriers to the left and to the right of the intermediate determines whether bond making or bond breaking is the rate determining step. In the diagram below, the formation of the M–Y bond has the highest free energy barrier and thus is rate determining.

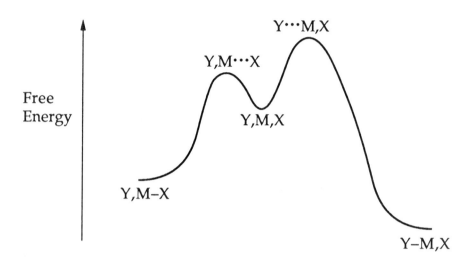

**15.4  Which experiment reveals the stoichiometric mechanism and which reveals the intimate mechanism? (a).** If the *trans* to *cis* isomerization of $[CoCl_2(en)_2]^+$ results in the incorporation of $^{36}Cl^-$ in the complex, then the isomerization reaction proceeded by a mechanism which includes the dissociation of $Cl^-$ from the octahedral complex. This is information that bears on the stoichiometric mechanism, i.e., the sequence of elementary steps by which the isomerization reaction takes place (see **Section 15.1**).

**(b).** On the other hand, if we find that the replacement of H by D in the ammine ligands of $[Cr(NCS)_4(NH_3)_2]^-$ reduces the rate of substitution of $NCS^-$ by $H_2O$, we have learned that N–H bond breaking is involved in the rate determining step, i.e., in the formation of the activated complex. This is a detail about the energetics of formation of the activated complex, so it bears on the intimate mechanism of substitution (base catalyzed hydrolysis, in this case).

**15.5  How would you determine if the formation of $[MnX(OH_2)_5]^+$ is $d$ or $a$?** The rate law for this substitution reaction is:

$$\text{rate} = (kK_E[Mn(OH_2)_6^{2+}][X^-])/(1 + K_E[X^-])$$

where $K_E$ is the equilibrium constant for the formation of the encounter complex $\{[Mn(OH)_6]^{2+}, X^-\}$, and $k$ is the first order rate constant for the reaction:

$$\{[Mn(OH)_6^{2+}], X^-\} \longrightarrow [Mn(OH)_5X]^+ + H_2O$$

The rate law will be the same regardless of whether the transformation of the encounter complex into products is dissociatively or associatively activated. However, we can distinguish $d$ from $a$ by varying $X^-$. If $k$ varies as $X^-$ varies, then the reaction is $a$. If $k$ is relatively constant as $X^-$ varies, then the reaction is $d$. Note that $k$ cannot be measured directly. It can be found using the expression $k_{obs} = kK_E$ and an estimate of $K_E$, as described in **Section 15.5**.

**15.6    The nonlability of octahedral complexes of metals with high oxidation numbers or metals of Periods 5 and 6.** If ligand substitution takes place by a $d$ mechanism, the strength of the metal-leaving group bond is directly related to the substitution rate. Metal centers with high oxidation numbers will have stronger bonds to ligands than metal centers with low oxidation numbers (see **Section 7.5**). Furthermore, Period 5 and 6 $d$-block metals have stronger metal ligand bonds (see also **Section 7.5**). Therefore, for reactions that are dissociatively activated, complexes of Period 5 and 6 metals are less labile than complexes of Period 3 metals, and complexes of metals in high oxidation states are less labile than complexes of metals in low oxidation states (all other things remaining equal).

**15.7    Explain in terms of associative activation the fact that a Pt(II) complex of tetraethyldiethylenetriamine is attacked by Cl$^-$ $10^5$ times less rapidly than the diethylenetriamine analog.** The two complexes are shown below. The ethyl substituted complex presents a greater degree of steric hindrance to an incoming Cl$^-$ ion nucleophile. Since the rate determining step for associative substitution of X by Cl$^-$ is the formation of a Pt–Cl bond, the more hindered complex will react more slowly.

**15.8    Explain why substitution reactions of $[Ni(CN)_4]^{2-}$ are very fast.** The fact that the five-coordinate complex $[Ni(CN)_5]^{3-}$ can be detected does indeed explain why substitution reactions of the four-coordinate complex $[Ni(CN)_4]^{2-}$ are fast. The reason is that, for a detectable amount of $[Ni(CN)_5]^{3-}$ to build up in solution, the forward rate constant $k_f$ must be numerically close to or greater than the reverse rate constant $k_r$:

$$[Ni(CN)_4]^{2-} + CN^- \underset{k_f}{\overset{k_r}{\rightleftharpoons}} [Ni(CN)_5]^{3-}$$

If $k_f$ were much smaller than $k_r$, the equilibrium constant $K = k_f/k_r$ would be small and the concentration of $[Ni(CN)_5]^{3-}$ would be too small to detect. Therefore, since $k_f$ is relatively large, we can infer that rate constants for the association of other nucleophiles are also large, with the result that substitution reactions of $[Ni(CN)_4]^{2-}$ are very fast.

**15.9    A two-step synthesis for *cis*- and *trans*-$[PtCl_2(NO_2)(NH_3)]^-$.** Starting with $[PtCl_4]^{2-}$, we need to perform two separate ligand substitution reactions. In one, $NH_3$ will replace $Cl^-$ ion. In the other, $NO_2^-$ ion will replace $Cl^-$ ion. The question is, which substitution to perform first. According to the *trans* effect series shown in **Section 15.3**, the strength of the *trans* effect on Pt(II) for the three ligands in question is $NH_3 < Cl^- < NO_2^-$. This means that a $Cl^-$ ion *trans* to another $Cl^-$ will be substituted faster than a $Cl^-$ ion *trans* to $NH_3$, while a $Cl^-$ ion *trans* to $NO_2^-$ will be substituted faster than a $Cl^-$ ion *trans* to another $Cl^-$ ion. If we first add $NH_3$ to $[PtCl_4]^{2-}$, we will produce

$[PtCl_3(NH_3)]^-$. Now if we add $NO_2^-$, one of the mutually *trans* $Cl^-$ ligands will be substituted faster than the $Cl^-$ ligand *trans* to $NH_3$, and the *cis* isomer will be the result. If we first add $NO_2^-$ to $[PtCl_4]^{2-}$, we will produce $[PtCl_3(NO_2)]^{2-}$. Now if we add $NH_3$, the $Cl^-$ ligand *trans* to $NO_2^-$ will be substituted faster than one of the mutually *trans* $Cl^-$ ligands, and the *trans* isomer will be the result. These two-step syntheses are shown below:

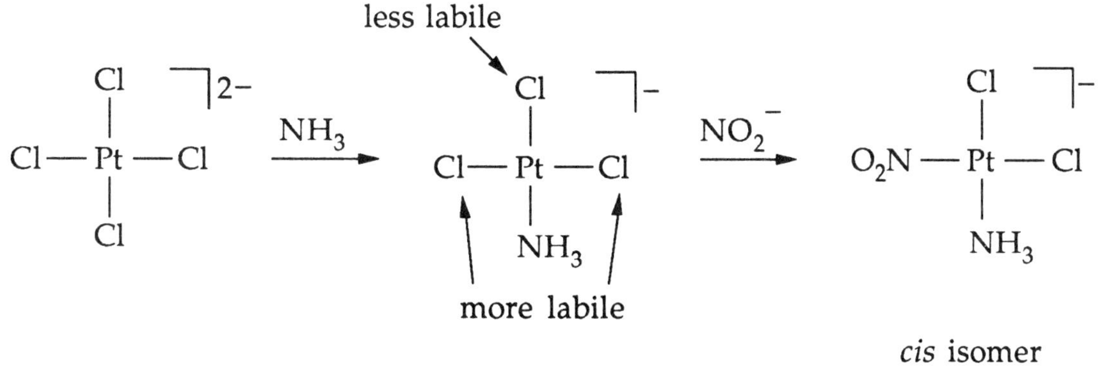

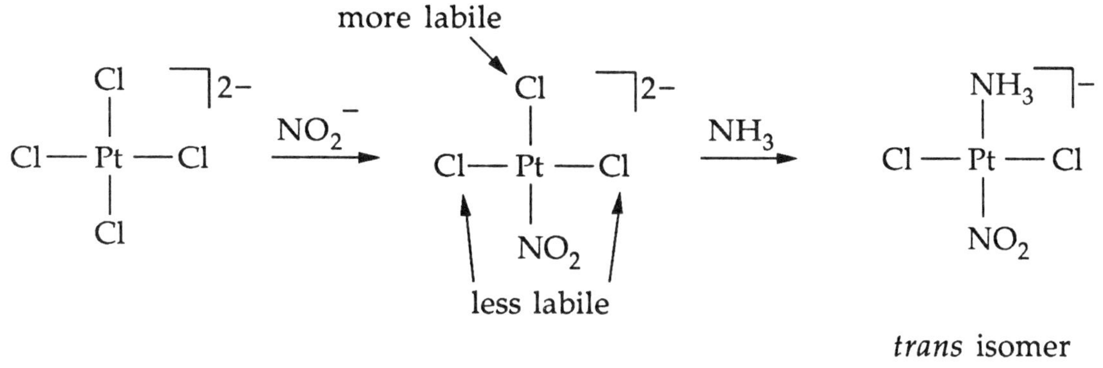

**15.10 How does each of the following affect the rate of square-planar substitution reactions? (a) Changing a *trans* ligand from H to Cl.** Hydride ion lies higher in the *trans* effect series than does chloride ion. Thus, if the ligand *trans* to H or Cl is the leaving group, its rate of substitution will be decreased if H is changed to Cl. The change in rate can be as large as a factor of $10^4$ (see **Table 15.4**).

(b) **Changing the leaving group from Cl to I.** The rate at which a ligand is substituted in a square-planar complex is related to its position in the *trans* effect series. If a ligand is high in the series, it is a *good* entering group (i.e., a good nucleophile) and a *poor* leaving group. Since iodide ion is higher in the *trans* effect series than chloride ion, it is a poorer leaving group than chloride ion. Therefore, the iodo complex will undergo I$^-$ substitution more slowly than the chloro complex will undergo Cl$^-$ substitution.

(c) **Adding a bulky substituent to a *cis* ligand.** This change will hinder the approach of the entering group and will slow the formation of the five-coordinate activated complex. The rate of substitution of a square-planar complex with bulky ligands will be slower than a comparable complex with sterically smaller ligands (see also **End-of-Chapter Exercise 15.7**).

(d) **Increasing the positive charge on the complex.** If all other things are kept equal, increasing the positive charge on a square-planar complex will increase the rate of substitution. This is because the entering ligands are either anions or the negative ends of dipoles. As explained in **Section 15.2**, if the charge density of the complex decreases in the activated complex, as would happen when an anionic ligand adds to a cationic complex, the solvent molecules will be *less* ordered around the complex (the opposite of the process called electrostriction). The increased disorder of the solvent makes $\Delta S^\ddagger$ less negative (compare the values of $\Delta S^\ddagger$ for the Pt(II) and Au(III) complexes in **Table 15.5**).

**15.11 Why is the rate of attack on Co(III) nearly independent of the entering group with the exception of OH$^-$?** The general trend is easy to explain: octahedral Co(III) complexes undergo dissociatively activated ligand substitution. The rate of substitution depends on the nature of the bond between the metal and the leaving group, since this bond is partially broken in the activated complex. The rate is independent of the nature of the bond to the entering group, since this bond is formed in a step subsequent to the rate determining step. The anomalously high rate of substitution by OH$^-$ signals an alternate path, that of base hydrolysis, as shown below (see **Section 15.8**):

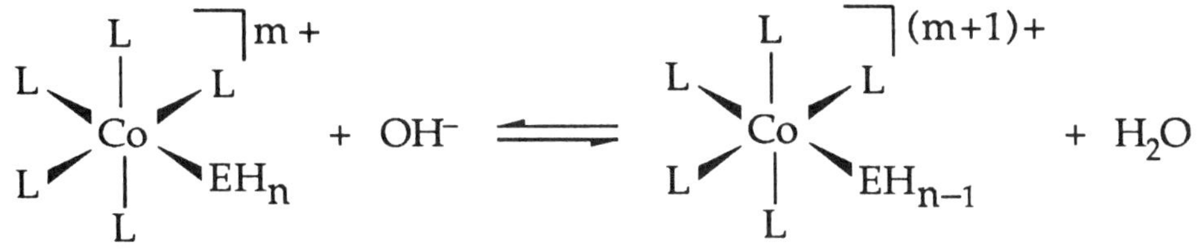

The deprotonated complex $[CoL_5(EH_{n-1})]^{(m+1)+}$ will undergo loss of L faster than the starting complex $[CoL_5(EH_n)]^{m+}$, because the anionic $EH_{n-1}{}^-$ ligand is a stronger base than the neutral $EH_n$ ligand and can better stabilize the coordinatively unsaturated activated complex. The implication is that a complex without protic ligands will not undergo anomalously fast $OH^-$ ion substitution.

**15.12 Predict the products of the following reactions: (a) $[Pt(PR_3)_4]^{2+} + 2Cl^-$.** The first $Cl^-$ ion substitution will produce the reaction intermediate $[Pt(PR_3)_3Cl]^+$. This will be attacked by the second equivalent of $Cl^-$ ion. Since phosphines are higher in the *trans* effect series than chloride ion, a phosphine *trans* to another phosphine will be substituted, giving *cis*-$[PtCl_2(PR_3)_2]$.

**(b) $[PtCl_4]^{2-} + 2PR_3$.** The first $PR_3$ substitution will produce the reaction intermediate $[PtCl_3(PR_3)]^-$. This will be attacked by the second equivalent of $PR_3$. Since phosphines are higher in the *trans* effect series than chloride ion, $Cl^-$ ion *trans* to another $Cl^-$ ion will be substituted, giving *trans*-$[PtCl_2(PR_3)_2]$.

**(c) $[Pt(py)_4]^{2+} + 2Cl^-$.** The first $Cl^-$ ion substitution will produce the reaction intermediate $[Pt(py)_3Cl]^+$. This will be attacked by the second equivalent of $Cl^-$ ion. Since $Cl^-$ ion is higher in the *trans* effect series than pyridine, a pyridine ligand *trans* to another pyridine ligand will be substituted, giving *trans*-$[PtCl_2(py)_2]$.

**15.13 Put the following in order of rate of substitution by $H_2O$: $[Co(NH_3)_6]^{3+}$, $[Rh(NH_3)_6]^{3+}$, $[Ir(NH_3)_6]^{3+}$, $[Mn(OH_2)_6]^{2+}$, $[Ni(OH_2)_6]^{2+}$.** The three ammine complexes will undergo substitution more slowly than the two aqua complexes. This is because of their higher charge and their low-spin $d^6$ configurations. You should refer to **Table 15.7**, which lists ligand-field activation energies (LFAE) for various $d^n$ configurations. While low-spin $d^6$ is not included, note the similarity between $d^3$ ($t_{2g}^3$) and low-spin $d^6$ ($t_{2g}^6$). A low-spin octahedral $d^6$ complex has LFAE = $0.4\Delta_o$, and consequently is inert to substitution. Of the three ammine complexes listed above, the iridium complex is the most inert, followed by the rhodium complex, which is more inert than the cobalt complex. This is because $\Delta_o$ increases on descending a group in the $d$-block. Of the two aqua complexes, the Ni complex, with LFAE = $0.2\Delta_o$, undergoes substitution more slowly than than the Mn complex, with LFAE = 0. Thus, the order of increasing rate is $[Ir(NH_3)_6]^{3+} < [Rh(NH_3)_6]^{3+} < [Co(NH_3)_6]^{3+} < [Ni(OH_2)_6]^{2+} < [Mn(OH_2)_6]^{2+}$.

**15.14 State the effect on the rate of dissociatively activated reactions of Rh(III) complexes of each of the following: (a) An increase in the positive charge on the complex.** Since the leaving group (X) is invariably negatively charged or the negative end of a dipole, increasing the positive charge on the complex will retard the rate of M–X bond cleavage. For a dissociatively activated reaction, this change will result in a decreased rate.

**(b) Changing the leaving group from $NO_3^-$ to $Cl^-$.** Try to answer this one after inspecting **Figure 15.4**. For the substitution reaction shown, changing the leaving group from nitrate ion to chloride ion results in a decreased rate. The explanation offered in **Section 15.6** is that the Co–Cl bond is stronger than the Co–$ONO_2$ bond. For a dissociatively activated reaction, a stronger bond to the leaving group will result in a decreased rate.

**(c) Changing the entering group from $Cl^-$ to $I^-$.** This change will have little or no effect on the rate. For a dissociatively activated reaction, the bond between the entering group and the metal is formed subsequent to the rate determining step.

**(d) Changing the _cis_ ligands from $NH_3$ to $H_2O$.** These two ligands differ in their σ-basicity. The more basic ligand, $NH_3$, will increase the electron density at the metal and will help stabilize the coordinatively unsaturated activated complex. Therefore, this change, from $NH_3$ to the less basic $H_2O$, will result in a decreased rate.

**(e) Changing an ethylenediamine ligand to propylenediamine when the leaving group is Cl⁻.** Ethylenediamine forms five-membered chelate rings with metal ions while propylenediamine forms six-membered chelate rings. The greater flexibility of the latter will lead to a more stabilized activated complex. Therefore, this change will lead to an increased rate.

**15.15 What data might distinguish between an inner- and outer-sphere pathway for reduction of $[Co(N_3)(NH_3)_5]^{2+}$ with $[V(OH_2)_6]^{2+}$?** The inner-sphere pathway is shown below (solvent = $H_2O$):

$$[Co(N_3)(NH_3)_5]^{2+} + [V(OH_2)_6]^{2+} \rightarrow \{[Co(N_3)(NH_3)_5]^{2+}, [V(OH_2)_6]^{2+}\}$$

$$\{[Co(N_3)(NH_3)_5]^{2+}, [V(OH_2)_6]^{2+}\} \rightarrow \{[Co(N_3)(NH_3)_5]^{2+}, [V(OH_2)_5]^{2+}, H_2O\}$$

$$\{[Co(N_3)(NH_3)_5]^{2+}, [V(OH_2)_5]^{2+}, H_2O\} \rightarrow [(NH_3)_5Co-N=N=N-V(OH_2)_5]^{4+}$$
$$\qquad\qquad\qquad\qquad\qquad\qquad\qquad\qquad\qquad\qquad Co(III) \qquad\qquad V(II)$$

$$[(NH_3)_5Co-N=N=N-V(OH_2)_5]^{4+} \rightarrow [(NH_3)_5Co-N=N=N-V(OH_2)_5]^{4+}$$
$$\qquad\quad Co(III) \qquad\quad V(II) \qquad\qquad\qquad\qquad\quad Co(II) \qquad\quad V(III)$$

$$[(NH_3)_5Co-N=N=N-V(OH_2)_2]^{4+} \rightarrow \rightarrow [Co(OH_2)_6]^{2+} + [V(N_3)(OH_2)_5]^{2+}$$
$$\qquad Co(II) \qquad\qquad V(III)$$

The pathway for outer-sphere electron transfer is shown below:

$$[Co(N_3)(NH_3)_5]^{2+} + [V(OH_2)_6]^{2+} \rightarrow \{[Co(N_3)(NH_3)_5]^{2+}, [V(OH_2)_6]^{2+}\}$$

$$\{[Co(N_3)(NH_3)_5]^{2+}, [V(OH_2)_6]^{2+}\} \rightarrow \{[Co(N_3)(NH_3)_5]^{+}, [V(OH_2)_6]^{3+}\}$$

$$\{[Co(N_3)(NH_3)_5]^{+}, [V(OH_2)_6]^{3+}\} \rightarrow [Co(OH_2)_6]^{2+} + [V(OH_2)_6]^{3+}$$

In both cases, the cobalt-containing product is the aqua complex, since $H_2O$ is present in abundance and high-spin $d^7$ complexes of Co(II) are substitution labile. However, something that distinguishes the two pathways is the composition of the vanadium-containing product. If $[V(N_3)(OH_2)_5]^{2+}$ is the product, then the reaction has proceeded *via* an inner-sphere pathway. If $[V(OH_2)_6]^{3+}$ is the product, then the electron transfer reaction is outer-sphere.

**15.16 The presence of detectable intermediates in electron transfer reactions.** As in **Exercise 15.15**, above, the direct transfer of a ligand from the coordination sphere of one redox partner (in this case the oxidizing agent, $[Co(NCS)(NH_3)]^{2+}$) to the coordination sphere of the other (in this case the reducing agent, $[Fe(OH_2)_6]^{3+}$), signals an inner-sphere electron transfer reaction. Even if the first formed product $[Fe(NCS)(OH_2)_5]^{2+}$ is short lived and undergoes hydrolysis to $[Fe(OH_2)_6]^{3+}$, its fleeting existence demands that the electron was transfered across a Co–(NCS)–Fe bridge.

**15.17 What product and quantum yield do you predict for substitution of $[W(CO)_5(py)]$ in the presence of excess triethylamine?** Since the intermediate is believed to be $[W(CO)_5]$, the properties of the entering group (triethylamine vs. triphenylphosphine) should not affect the quantum yield of the reaction, which is a measure of the rate of formation of $[W(CO)_5]$ from the excited state of $[W(CO)_5(py)]$. The product of the photolysis of $[W(CO)_5(py)]$ in the presence of excess triethylamine will be $[W(CO)_5(NEt_3)]$, and the quantum yield will be 0.4. This photosubstitution is initiated from a ligand-field excited state, not a MLCT excited state. A metal-ligand charge transfer increases the oxidation state of the metal, which would strengthen, not weaken, the bond between the metal and a σ-base like pyridine.

**15.18 Propose a wavelength for photoreduction of $[CoCl(NH_3)_5]^{2+}$.** The intense band at ~250 nm in the spectrum shown in **Figure 14.6** cannot be a Laporte-forbidden ligand-field transition. It is most likely a LMCT transition (specifically a $Cl^-$-to-$Co^{3+}$ charge transfer). Irradiation at this wavelength should produce a population of $[CoCl(NH_3)_5]^{2+}$ ions that contain $Co^{2+}$ ions and Cl atoms instead of $Co^{3+}$ ions and $Cl^-$ ions. If these excited state complexes live long enough (i.e., if they do not rapidly decay to the ground state), separation of the $Co^{2+}$ ions and Cl atoms will result in the reduction of $[CoCl(NH_3)_5]^{2+}$. Irradiation of the complex at wavelengths between 350 and 600 nm will not lead to photoreduction. The band that are observed between these two wavelengths are ligand-field transitions: the electrons on the metal are rearranged, and the electrons on the $Cl^-$ ion are not involved.

# CHAPTER 16

# *d-* AND *f-*BLOCK ORGANOMETALLICS

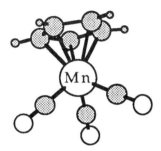

The structure of tricarbonyl($\eta^5$-cyclopentadienyl)-manganese(I) (the oxygen atoms of the carbonyl ligands are not shaded). This molecule contains two important types of ligands frequently found in organometallic compounds, $\pi$-acids (the CO ligands) and unsaturated organic molecules or molecular fragments that donate their $\pi$ electrons to the metal center (the $\eta^5$-$C_5H_5$ ligand).

## SOLUTIONS TO IN-CHAPTER EXERCISES

**16.1 Is Mo(CO)$_7$ likely to exist?** A Mo atom (Group 6) has six valence electrons, and each CO ligand is a two-electron donor. Therefore, the total number of valence electrons on the Mo atom in this compound would be $6 + 7(2) = 20$. Since organometallic compounds with more than 18 valence electrons on the central metal are never stable, Mo(CO)$_7$ is *not* likely to exist. The compound Mo(CO)$_6$, with exactly 18 valence electrons, is very stable. Note that throughout **Chapter 16**, the authors do not always write the formulas for organometallic complexes in square brackets. According to the rules of inorganic nomenclature, *all* complexes should be written in square brackets. Nevertheless, the authors are following a common, informal set of rules: neutral organometallic complexes are written without square brackets, while cationic or anionic organometallic complexes are written with them.

**16.2  Assign the oxidation number of Co in Co($\eta^5$-C$_5$H$_5$)(CO)$_2$.** The CO ligands are neutral two-electron donors, so their oxidation number is 0. The pentahaptocyclopentadienyl ligand is assigned an oxidation number −1. The complex as a whole is neutral, so the sum of ligand oxidation numbers (−1 + 0 + 0 = −1) must be equal in magnitude but opposite in sign to the metal oxidation number. Therefore, the oxidation number of Co in this compound is +1.

**16.3  Does Ni$_2$Cp$_2$(CO)$_2$ contain bridging or terminal CO ligands or both?** The CO stretching bands at 1857 cm$^{-1}$ and 1897 cm$^{-1}$ are both lower in frequency than typical terminal CO ligands (for terminal CO ligands, $v$(CO) > 1900 cm$^{-1}$ (see **Figure 16.2**). Therefore, it seems likely that it only contains bridging CO ligands. The presence of two bands suggests that the bridging CO ligands are probably *not* collinear, since only one band would be observed it they were (see **Figure 16.5**).

**16.4  If Mo(CO)$_3$L$_3$ is desired, which of the ligands P(CH$_3$)$_3$ or P($i$-Bu)$_3$ would be preferred?** Since this is a highly substituted complex, the effects of steric crowding must be considered. This is especially true in this case, since the two ligands in question should be very similar electronically. The cone angle for P(CH$_3$)$_3$, given in **Table 16.6**, is 118°. While the cone angle for P($i$-Bu)$_3$ is not given in the table, we can surmise that it is considerably larger than 118° (for example, the cone angle of PEt$_3$ is 137°). Therefore, because of its smaller size, PMe$_3$ would be preferred.

**16.5  Propose a synthesis for Mn(CO)$_4$(PPh$_3$)(COCH$_3$).** Consider the reactions of carbonyl complexes discussed in **Section 16.6**. If we use Mn$_2$(CO)$_{10}$ as the source of manganese, we can reductively cleave the Mn–Mn bond with sodium, forming Na[Mn(CO)$_5$]:

$$\text{Mn}_2(\text{CO})_{10} + 2\text{Na} \rightarrow 2\text{Na}[\text{Mn}(\text{CO})_5]$$

oxidation number:     0                        0           +1  −1

The anionic carbonyl complex is a relatively good nucleophile. When it is treated with $CH_3I$, it displaces $I^-$ to form $Mn(CH_3)(CO)_5$ (sometimes written as $CH_3Mn(CO)_5$):

$$Na[Mn(CO)_5] + CH_3I \rightarrow Mn(CH_3)(CO)_5 + NaI$$

Many alkyl-substituted metal carbonyls undergo a migratory insertion reaction when treated with basic ligands. The alkyl group (methyl in this case) migrates from the Mn atom to an adjacent C atom of a CO ligand, leaving an open coordination site for the entering group ($PPh_3$ in this case) to attack:

$$Mn(CH_3)(CO)_5 + PPh_3 \rightarrow Mn(CO)_4(PPh_3)(COCH_3)$$

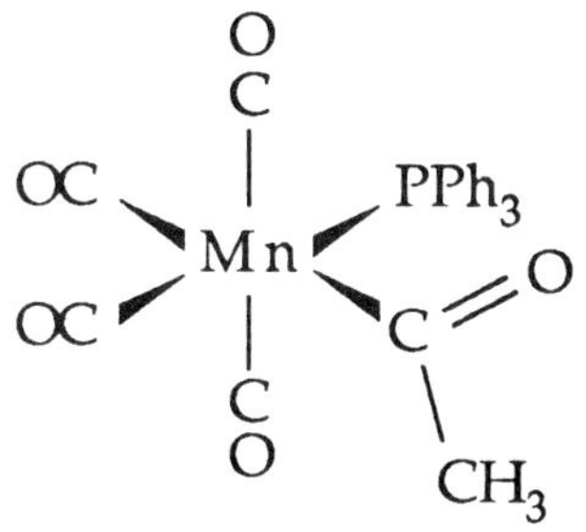

The structure of $Mn(CO)_4(PPh_3)(COCH_3)$

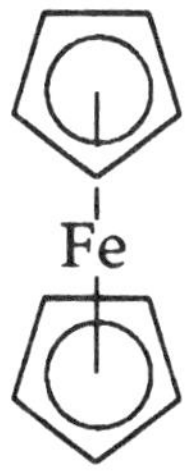

The structure of ferrocene

## 16.6 Will oxidation of $FeCp_2$ to $[FeCp_2]^+$ produce a substantial change in M–C bond length?

Neutral ferrocene contains 18 valence electrons (8 from Fe and 10 from the two Cp ligands). The MO diagram for ferrocene is shown in **Figure 16.12**. Eighteen electrons will fill it up to the second $a_1'$ orbital. Since this orbital is the HOMO, oxidation of ferrocene will result in removal of an electron from it, leaving the 17 electron ferricenium cation, $[FeCp_2]^+$. If this orbital were strongly bonding, removal of an electron would result in weaker Fe–C bonds. If this orbital were strongly antibonding, removal of an electron would result in stronger Fe–C bonds. However, this orbital is essentially nonbonding (see **Figure 16.13**). Therefore, oxidation of $FeCp_2$ to $[FeCp_2]^+$ will not produce a substantial change in the Fe–C bond order or the Fe–C bond length.

**16.7    Propose a structure for Cp$_4$Fe$_4$(CO)$_4$.** There are four relevant pieces of information given. First of all, the fact that the compound is highly colored suggests that it contains metal-metal bonds. Second, the composition can be used to determine the cluster valence electron count, which can be used to predict which polyhedral structure is likely:

$$
\begin{array}{llll}
\text{Fe}_4 & 4 \times 8\ e^- = & 32\ \text{valence}\ e^- \\
\text{Cp}_4 & 4 \times 5 = & 20 \\
(\text{CO})_4 & 4 \times 2 = & 8 \\
\hline
\text{Total} = & & 60\ \text{valence}\ e^- \Rightarrow \text{a tetrahedral cluster}
\end{array}
$$

Third, the presence of only one line in the $^1$H NMR spectrum suggests that the Cp ligands are all equivalent and are $\eta^5$. Finally, the single CO stretch at 1640 cm$^{-1}$ suggests that the CO ligands are triply bridging (see **Figure 16.2**) and that they form a relatively high symmetry array (otherwise there would be more bands — see **Table 16.5**). A likely structure for Fe$_4$Cp$_4$(CO)$_4$ is:

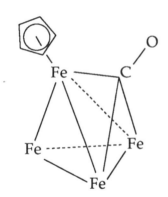

The tetrahedral Fe$_4$ core of this cluster exhibits six equivalent Fe–Fe bond distances. Only one of the $\eta^5$-C$_5$H$_5$ ligands and one of the triply bridging CO ligands is shown. Each Fe atom has a Cp ligand, and each of the four triangular faces of the Fe$_4$ tetrahedron is capped by a triply bridging CO ligand.

---

## SOLUTIONS TO END-OF-CHAPTER EXERCISES

**16.1    Name and draw the structures of the following: (a) Fe(CO)$_5$.** The names of metal carbonyls first specify the number of CO ligands, then the number and type of metal atom(s), and finally the oxidation number of the metal(s). Therefore, the name of Fe(CO)$_5$ is pentacarbonyliron(0). As discussed in **Section 16.5**, the structures of metal carbonyls generally have simple, symmetrical shapes that correspond to the CO ligands taking up positions that place them as far apart from one another as possible. In this case the analogy is between an Fe atom with five CO ligands and an atom with five bonding pairs of valence electrons, like the P atom in $PF_5$. Thus, the structure of $Fe(CO)_5$ is a trigonal bipyramid, like $PF_5$, as shown below.

**(b)  Ni(CO)$_4$.** The name of this complex is tetracarbonylnickel(0). In this case the analogy is between a Ni atom with four CO ligands and an atom with four bonding pairs of valence electrons, like the C atom in $CH_4$. Thus, the structure of $Ni(CO)_4$ is tetrahedral, as shown below.

**(c)   Mo(CO)$_6$.**   The name of this complex is hexacarbonyl-molybdenum(0). The analogy here is between a Mo atom with six CO ligands and an atom with six bonding pairs of valence electrons, like the S atom in $SF_6$. Thus, the structure of $Mo(CO)_6$ is octahedral, as shown below.

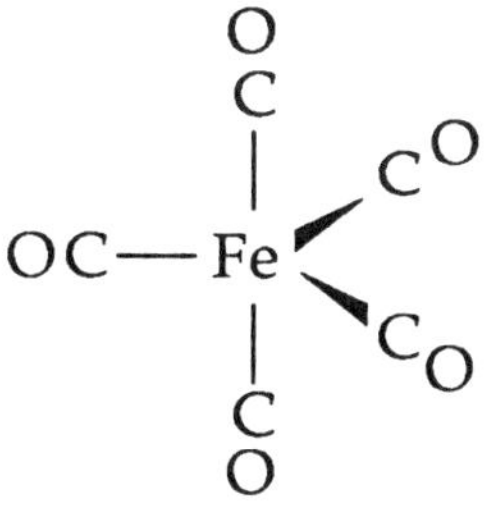

$$Fe(CO)_5 \qquad\qquad Ni(CO)_4 \qquad\qquad Mo(CO)_6$$

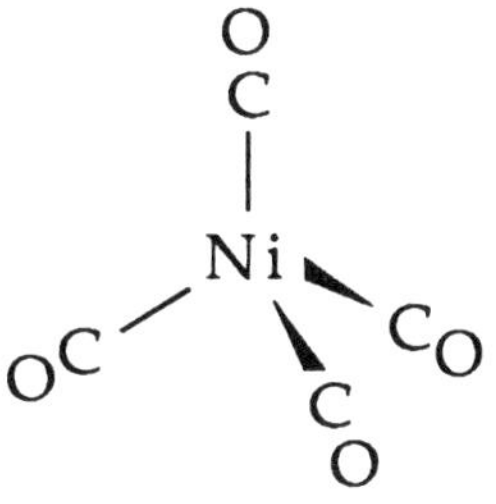

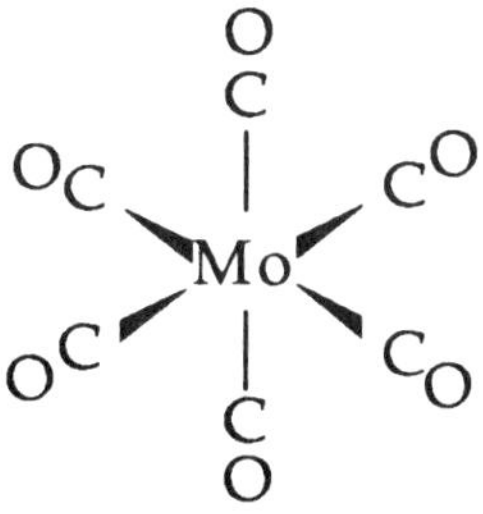

**(d)  Mn$_2$(CO)$_{10}$.** The name of this *dinuclear* (i.e., two metal atoms) complex is decacarbonyldimanganese(0). Its structure is based on two square-pyramidal $Mn(CO)_5$ fragments joined by a Mn–Mn bond. The $Mn(CO)_5$ fragments are staggered with respect to each other. The geometry around each Mn atom is octahedral, although the complex itself has $D_{4d}$ symmetry, not $O_h$ symmetry. Its structure is shown below.

**(e)  V(CO)$_6$.** The name of this complex is hexacarbonylvanadium(0). Its structure is the same as that of $Mo(CO)_6$, octahedral, as shown below.

**(f) [PtCl₃(C₂H₄)]⁻.** The name of this complex, which has two ligand types, is trichloro(ethylene)platinate(II) or trichloro(ethylene)platinate(1−). You should review the nomenclature rules discussed in **Section 7.2**. Complexes are named with their ligands in alphabetical order. The prefixes di-, tri-, tetra-, etc. do not count as far as the alphabetical order is concerned. Thus, *c* before *e* in trichloro(*e*thylene)platinate(II). The structure of this complex is square-planar, the usual structure for Period 4 and 5 $d^8$ metal ions, and is shown below.

$V(CO)_6$                                           $Mn_2(CO)_{10}$

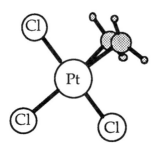

The structure of $[PtCl_3(C_2H_4)]^−$. The geometry around the Pt atom is square-planar, although the complex anion is not planar (the ethylene ligand is perpendicular to the PtCl₃ plane). Due to back-donation, the C=C bond distance in the complex, 1.375 Å, is slightly longer than the C=C bond distance in ethylene, 1.337 Å.

**16.2 Assign oxidation numbers in the following: (a) $[(\eta^5\text{-}C_5H_5)_2Fe][BF_4]$.** The tetrafluoroborate anion has a −1 charge, so the iron complex must have a +1 charge. Since each cyclopentadienyl ligand is generally considered to be a −1 anion, the oxidation number of iron is +3.

**(b) Fe(CO)₅.** The complex is not charged, and since each carbonyl ligand is neutral, the oxidation number of iron is 0.

(c) **[Fe(CO)₄]²⁻.** The complex has a –2 charge. Since each carbonyl ligand is neutral, the oxidation number of iron is –2.

(d) **Co₂(CO)₈.** The complex is not charged, and since each carbonyl ligand is neutral, the oxidation number of cobalt is 0.

**16.3  Do the following deviate from the 18-electron rule? Fe(CO)₅.** A Fe atom (Group 8) has eight valence electrons, and each CO ligand is a two-electron donor. Therefore, the total number of valence electrons on the Fe atom in this compound is $8 + 5(2) = 18$.

**Ni(CO)₄.** A Ni atom (Group 10) has ten valence electrons, and each CO ligand is a two-electron donor. Therefore, the total number of valence electrons on the Ni atom in this compound is $10 + 4(2) = 18$.

**Mo(CO)₆.** A Mo atom (Group 6) has six valence electrons, and each CO ligand is a two-electron donor. Therefore, the total number of valence electrons on the Mo atom in this compound is $6 + 6(2) = 18$.

**Mn₂(CO)₁₀.** A Mn atom (Group 7) has seven valence electrons, and each CO is a two-electron donor. Polynuclear metal carbonyls typically have metal-metal bonds, and this compound is no exception. The Mn–Mn bond allows each Mn atom to share an additional valence electron. Therefore, the total number of valence electrons on the Mn atom in this compound is $7 + 5(2) + 1 = 18$.

**V(CO)₆.** A V atom (Group 5) has five valence electrons, and each CO ligand is a two-electron donor. Therefore, the total number of valence electrons on the V atom in this compound is $5 + 6(2) = 17$. This octahedral complex deviates from the 18-electron rule. It is much more reactive than the first three metal carbonyls, undergoing ligand substitution reactions at very rapid rates. For example, the relative rates for M = V and Cr for the following reaction:

$$M(CO)_6 + PR_3 \rightarrow M(CO)_5(PR_3) + CO$$

are in the ratio of ~$10^{10}$. V(CO)$_6$ is readily reduced to [V(CO)$_6$]$^-$, an 18-electron complex anion. It is also *very* highly colored, in sharp contrast to Fe(CO)$_5$, Ni(CO)$_4$, and Mo(CO)$_6$, which are either colorless or only very faintly colored.

**[PtCl$_3$(C$_2$H$_4$)]$^-$.** The Pt atom (Group 10) has ten valence electrons, each Cl atom is a one-electron donor, ethylene is a two-electron donor, and one electron must be added for the –1 charge of the complex. Therefore, the total number of valence electrons on the Pt atom in this complex is 10 + 3(1) + 2 + 1 = 16. This complex deviates from the 18-electron, as do many four-coordinate Period 4 and 5 $d^8$ complexes (see **Section 16.1**). As a consequence of being two electrons short of 18, this complex undergoes ligand substitution by an associative mechanism (i.e., with the formation of a five-coordinate 18-electron intermediate).

**[($\eta^5$-C$_5$H$_5$)$_2$Fe]$^+$.** A Fe atom has eight valence electrons, each $\eta^5$-cyclo-pentadienyl ligand is a five-electron donor, and one electron must be subtracted for the +1 charge of the complex. Therefore, the total number of valence electrons on the Fe atom in this complex is 8 + 2(5) – 1 = 17. Like V(CO)$_6$, the deviation from the 18-electron rule causes this compound to be readily reduced by one electron (i.e., it is a strong oxidant).

**[Fe(CO)$_4$]$^{2-}$.** A Fe atom has eight valence electrons, each carbonyl ligand is a two electron donor, and two electrons must be added for the –2 charge on the complex. Therefore, the total number of valence electrons on the Fe atom in this complex is 8 + 2(2) + 2 = 18.

**Co$_2$(CO)$_8$.** Each Co atom (Group 9) has nine valence electrons, and each carbonyl is a two electron donor. The Co–Co bond allows each Co atom to share an additional valence electron. Therefore, since the total number of valence electrons of the *two* Co atoms in this compound is 2(9) + 8(2) + 2 = 36, each Co atom obeys the 18-electron rule. In the solid-state, this compound has two bridging carbonyl ligands and six terminal carbonyl ligands, but in solution it has only terminal carbonyl ligands, as shown below:

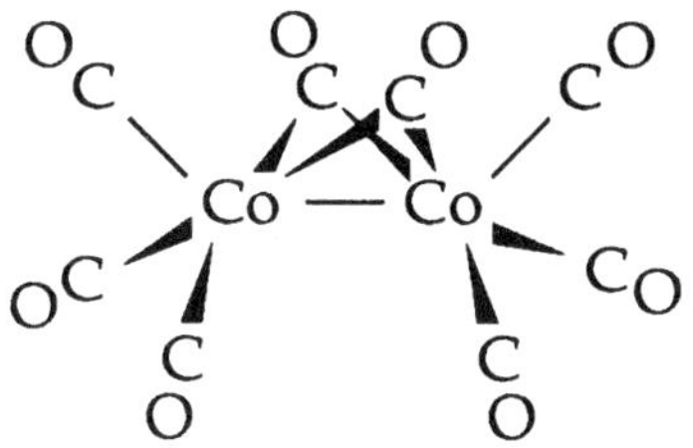

$Co_2(CO)_8$ in the solid–state

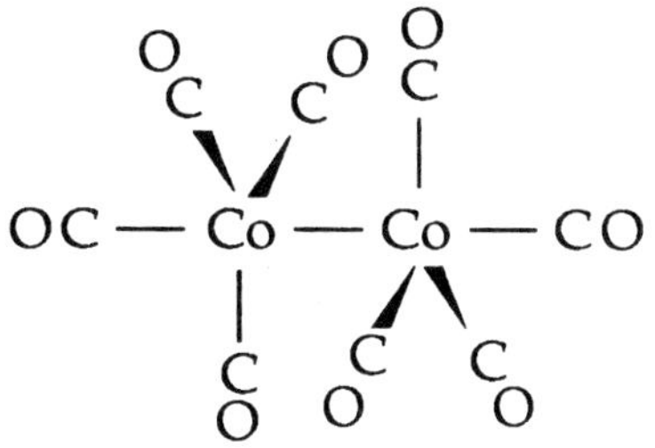

$Co_2(CO)_8$ in solution

**16.4    The common methods for the preparation of simple metal carbonyls.** As discussed in **Section 16.4**, the two principal methods are (1) direct combination of CO with a finely divided metal and (2) reduction of a metal salt in the presence of CO under pressure. Two examples are shown below, the preparation of hexacarbonylmolybdenum(0) and octacarbonyldicobalt(0). Other examples are given in the text.

(1)   $Mo\ (s) + 6CO\ (g) \rightarrow Mo(CO)_6\ (s)$   (high temp. and pressure required)

(2)   $2CoCO_3\ (s) + 2H_2\ (g) + 8CO\ (g) \rightarrow Co_2(CO)_8\ (s) + 2CO_2 + 2H_2O$

The reason that the second method is preferred is kinetic, not thermodynamic. The atomization energy (i.e., sublimation energy) of most metals is simply too high for the first method to proceed at a practical rate.

**16.5    A sequence of reactions for the preparation of $Fe(CO)_3$(diphos).** The most general way to prepare ligand substituted metal carbonyl complexes is to treat the parent binary metal carbonyl, in this case $Fe(CO)_5$, with the ligand of choice, in this case diphos (1,2-bis(diphenylphosphinoethane)). The two step reaction sequence is:

$Fe\ (s) + 5CO\ (g) \rightarrow Fe(CO)_5\ (l)$   (high temp. and pressure required)

$Fe(CO)_5\ (l) + diphos \rightarrow Fe(CO)_3(diphos)\ (s) + 2CO\ (g)$

The second step would require a slightly elevated temperature. A convenient way to achieve this would be to perform the ligand substitution in a refluxing solvent such as THF.

**16.6    IR spectra of metal tricarbonyl complexes.** In general, the lower the symmetry of a $M(CO)_n$ fragment, the greater the number of CO stretching bands in the IR spectrum. Therefore, given complexes with $M(CO)_3$ fragments that have either $C_{3v}$, $D_{3h}$, or $C_s$ symmetry, the complex that has $C_s$ symmetry will have the greatest number of bands. When you consult **Table 16.5**, you will see that a $M(CO)_3$ fragment with $D_{3h}$ symmetry will give rise to one band, one with $C_{3v}$ symmetry will give rise to two bands, and one with $C_s$ symmetry will give rise to three bands.

**16.7    $(C_5H_5)_3Ni_3(CO)_2$. (a) Propose a structure based on IR data.** The structure on the right is consistent with the IR spectroscopic data. This $D_{3h}$ complex has all three $\eta^5$-$C_5H_5$ ligands in identical environments. Furthermore, its has two collinear bridging CO ligands, which fits the single CO stretching band at a relatively low frequency. Although it was not given, you should expect this complex to be highly colored, due to the presence of metal-metal bonds.

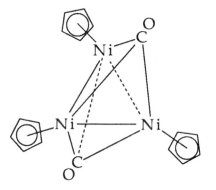

**(b)  Does each Ni atom obey the 18-electron rule?** The three Ni atoms are in identical environments, so we only have to determine the number of valence electrons for one of them:

| | |
|---|---|
| Ni | 10 valence $e^-$ |
| $\eta^5$-$C_5H_5$ | 5 |
| 2 Ni–Ni | 2 |
| 1/3(2 CO) | 4/3 |
| | ___ |
| Total | 18-1/3 |

Therefore, the Ni atoms in this trinuclear complex do not obey the 18-electron rule. Deviations from the rule are common for cyclopentadienyl complexes to the right of the *d*-block. For example, the stable complex $(\eta^5\text{-}C_5H_5)_2Co$ is a 19-electron compound.

**16.8   Which of the two complexes $W(CO)_6$ or $IrCl(CO)(PPh_3)_2$ should undergo the faster exchange with $^{13}CO$.** The 18-electron tungsten complex undergoes ligand substitution by a dissociative mechanism. The rate determining step involves cleavage of a relatively strong W–CO bond. In contrast, the 16-electron iridium complex undergoes ligand substitution by an associative mechanism, which does not involve Ir–CO bond cleavage in the activated complex. Accordingly, $IrCl(CO)(PPh_3)_2$ undergoes faster exchange with $^{13}CO$ than does $W(CO)_6$.

**16.9   Which complex should be more basic toward a proton? (a) $[Fe(CO)_4]^{2-}$ or $[Co(CO)_4]^-$.** The dianionic complex $[Fe(CO)_4]^{2-}$ should be the more basic. The trend involved is the greater affinity for a cation that a species with a higher negative will have, all other things equal. In this case the "other things" are (1) same set of ligands, (2) same structure (tetrahedral), and (3) same electron configuration ($d^{10}$). For a more detailed explanation, see the sub-section on **Metal Basicity** in **Section 16.6**.

**(b)   $[Mn(CO)_5]^-$ or $[Re(CO)_5]^-$.** The rhenium complex should be the more basic. The trend involved is the greater M–H bond enthalpy for a Period 6 metal ion relative to a period 4 metal ion in the same group, all other things equal. In this case the "other things" are (1) same set of ligands, (2) same structure (trigonal bipyramidal), and (3) same metal oxidation number (–1). Remember that in the *d*-block, bond enthalpies such as M–M, M–H, and M–R *increase* down a group. This behavior is opposite to that exhibited by the *p*-block elements (for example, the order of bond enthalpies for E–H bonds to Group 6 elements is O–H > S–H > Se–H > Te–H).

**16.10  What hapticities are possible for the following ligands: (a) $C_2H_4$.** Ethylene coordinates to *d*-block metals in only one way, using its $\pi$ electrons to form a metal-ethylene $\sigma$ bond (there may also be a significant amount of back donation, if ethylene is substituted with electron withdrawing groups; see **Section 16.7**). Therefore, $C_2H_4$ is always $\eta^2$, as shown below.

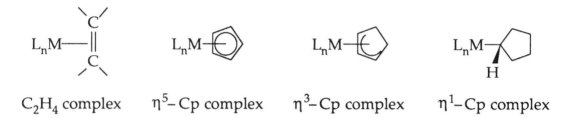

C₂H₄ complex     $\eta^5$– Cp complex     $\eta^3$– Cp complex     $\eta^1$– Cp complex

(b) **Cyclopentadienyl.** This is a very versatile ligand that can be $\eta^5$ (a five-electron donor), $\eta^3$ (a three-electron donor similar to simple allyl ligands; see **Structural Drawing 8**), or $\eta^1$ (a one-electron donor similar to simple alkyl and aryl ligands; see **Structural Drawing 7**). These three bonding modes are shown above (see **Section 16.8**)

(c) **$C_6H_6$.** This is also a versatile ligand, which can form $\eta^6$, $\eta^4$, and $\eta^2$ complexes. In such complexes, the ligands are, respectively, six-, four-, and two-electron donors. These three bonding modes are shown below.

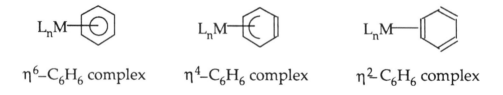

$\eta^6$–$C_6H_6$ complex     $\eta^4$–$C_6H_6$ complex     $\eta^2$–$C_6H_6$ complex

(d) **Butadiene.** This ligand can form both $\eta^4$ and $\eta^2$ complexes, in which they are four- and two-electron donors, respectively. A drawing of an $\eta^4$-butadiene complex is shown in **Section 16.7**. An $\eta^2$-butadiene complex would resemble an $\eta^2$-ethylene complex, except that one of the two C=C double bonds would remain uncoordinated.

(e) **Cyclooctatetraene.** This ligand contains four C=C double bonds, any combination of which can coordinate to a $d$-block (or $f$-block) metal. Thus *cyclo*-$C_8H_8$ can be $\eta^8$ (an eight electron donor; see **Structural Drawing 5**), $\eta^6$ (a six-electron donor), $\eta^4$ (a four-electron donor; see **Structural Drawings 42, 43**, and **44** and **Figures 16.14** and **16.15**), and $\eta^2$ (a two-electron donor, in which it would resemble an $\eta^2$-ethylene complex, except that three of the four C=C double bonds would remain uncoordinated).

**16.11 Draw structures and give the electron count of the following:**
**(a) $(\eta^3$-C$_3$H$_5)_2$Ni.** Bis(allyl)nickel(0) has the structure shown below. Each allyl ligand, which is planar, is a three-electron donor, so the number of valance electrons around the Ni atom (Group 10) is 10 + 2(3) = 16. Sixteen-electron complexes are very common for Group 9 and Group 10 elements, especially for Rh$^+$, Ir$^+$, Ni$^{2+}$, Pd$^{2+}$, and Pt$^{2+}$ (all of which are $d^8$).

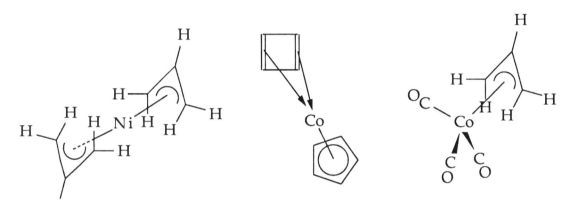

bis(allyl)nickel(0)     (cyclobutadiene)CpCo     (allyl)tricarbonylcobalt(0)

(b) $(\eta^4$-C$_4$H$_4)(\eta^5$-C$_5$H$_5)$Co. This complex has the structure shown above. Since the $\eta^4$-C$_4$H$_4$ ligand is a four-electron donor and the $\eta^5$-C$_5$H$_5$ ligand is a five-electron donor, the number of valance electrons around the Co atom (Group 9) is 9 + 4 + 5 = 18.

(c) $(\eta^3$-C$_3$H$_5)$Co(CO)$_3$. This complex has the structure shown above. The electron count for the Co atom is:

| | |
|---|---|
| Co | 9 valence $e^-$ |
| $\eta^3$-C$_3$H$_5$ | 3 |
| 3CO | 6 |
| | —— |
| Total | 18 $e^-$ |

**16.12 The *d*-block of the periodic table. (a) Which elements form neutral 18-electron Cp$_2$M compounds?** Since $\eta^5$-Cp is a five-electron donor, the only metals that form such compounds have 8 valence electrons. These are the Group 8 metals, Fe, Ru, and Os.

**(b) Which Period 4 elements form dimeric binary carbonyls?** The binary metal carbonyls for Period 4 elements are V(CO)$_6$, Cr(CO)$_6$, Mn$_2$(CO)$_{10}$, Fe(CO)$_5$, Co$_2$(CO)$_8$, and Ni(CO)$_4$ (see **Table 16.2**). The Mn and Co complexes are the only dimeric ones.

**(c) Which Period 4 elements form neutral carbonyls with six, five, and four carbonyl ligands?** From the list of neutral binary carbonyls given in part **(b)**, above, the elements in question are Cr, Fe, and Ni, respectively.

**(d) Which elements have the greatest tendency to follow the 18-electron rule?** The elements in the middle of the *d*-block have the greatest tendency. These are the chromium group (Cr, Mo, and W), the manganese group (Mn, Tc, and Re), and the iron group (Fe, Ru, and Os).

**16.13 Indicate the number of CO ligands in the following complexes: (a) ($\eta^6$-C$_6$H$_6$)W(CO)$_n$.** A W atom (Group 6) has six valence electrons, an $\eta^6$-C$_6$H$_6$ ligand is a six-electron donor, and each carbonyl ligand is a two-electron donor. Therefore, to satisfy the 18-electron rule, n = 3:

$$18 = 6 + 6 + 2n$$

**(b) ($\eta^5$-Cp)Rh(CO)$_n$.** A Rh atom (group 9) has nine valence electrons, an $\eta^5$-C$_5$H$_5$ ligand is a five-electron donor, and each carbonyl ligand is a two-electron donor. Therefore, to satisfy the 18-electron rule, n = 2:

$$18 = 9 + 5 + 2n$$

**(c) Ru$_3$(CO)$_n$.** This complex could be linear (i.e., Ru–Ru–Ru), with two Ru–Ru bonds, or triangular, with three Ru–Ru bonds. If it was linear, there would be no way to satisfy the 18-electron rule for the two Ru atoms (Group 8) at the end, regardless of the value of n:

$$\text{for all n,} \quad 18 \neq 8 + 1 \text{ (for Ru–Ru bond)} + 2n$$

Therefore, this must be a triangular cluster, with two Ru–Ru bonds per Ru atom. Therefore, m = 4 and n = 3m = 12, so the complex must be $Ru_3(CO)_{12}$:

$$18 = 8 + 2 \text{ (for two Ru–Ru bonds)} + 2m$$

**16.14 Give plausible equations for the synthesis of M–C, M–H, and M–M' bonds using metal carbonyl anions.** Metal carbonyl anions are Lewis bases, and as such can displace weaker bases from their complexes with Lewis acids such as $H^+$, $R^+$, and $L_nM^+$. Examples are shown in the following equations:

$$(\eta^5\text{-Cp})W(CO)_3^- + HCl \rightarrow (\eta^5\text{-Cp})W(CO)_3H + Cl^-$$

$$(\eta^5\text{-Cp})Fe(CO)_2^- + C_2H_5Cl \rightarrow (\eta^5\text{-Cp})Fe(CO)_2(C_2H_5) + Cl^-$$

$$(\eta^5\text{-Cp})W(CO)_3^- + (\eta^5\text{-Cp})Fe(CO)_2Cl \rightarrow (\eta^5\text{-Cp})(CO)_3W\text{–}Fe(\eta^5\text{-Cp})(CO)_2 + Cl^-$$

**16.15 Propose a synthesis for $HMn(CO)_5$.** The most general way to prepare metal carbonyl hydrides is to protonate metal carbonyl anions. In this case, the single protonation of $Mn(CO)_5^-$ would yield the desired product. The anion could be produce by the reductive cleavage of $Mn_2(CO)_{10}$ with Na. Since sodium is a solid that is not soluble in typical solvents such as THF, a liquid amalgam of sodium with mercury is generally used. The mixture of two liquid phases, a THF solution of $Mn_2(CO)_{10}$ and Na(Hg), react more rapidly than a mixture of a THF solution and solid Na:

$$Mn_2(CO)_{10} \text{ (soln)} + 2Na(Hg) \text{ (l)} \rightarrow 2Na[Mn(CO)_5] \text{ (soln)} + Hg \text{ (l)}$$

$$Na[Mn(CO)_5] \text{ (s)} + H_3PO_4 \text{ (l)} \rightarrow HMn(CO)_5 \text{ (s)} + NaH_2PO_4 \text{ (s)}$$

The two solid products of the second reaction can be easily separated by sublimation, since $HMn(CO)_5$, like many metal carbonyls and metal carbonyl hydrides, is a volatile compound. The reason that phosphoric acid is preferred is that it is not oxidizing (metal carbonyl anions are notoriously prone to oxidation).

**16.16 The product obtained when Mo(CO)₆ is treated with PhLi and then with CH₃OSO₂CF₃.** The product of the first reaction contains a –C(=O)Ph ligand formed by the nucleophilic attack of Ph⁻ on one of the carbonyl C atoms:

$$Mo(CO)_6 + PhLi \rightarrow Li[Mo(CO)_5(COPh)]$$

The most basic site on this anion is the acyl oxygen atom, and it is the site of attack by the methylating agent CH₃OSO₂CF₃:

$$Li[Mo(CO)_5(COPh)] + CH_3OSO_2CF_3 \rightarrow Mo(CO)_5(C(OCH_3)Ph) + LiSO_3CF_3$$

The final product is a carbene (alkylidene) complex. Since the carbene C atom has an oxygen-containing substituent, it is an example of a Fischer carbene (see **Section 16.7**). The two reaction products are shown below:

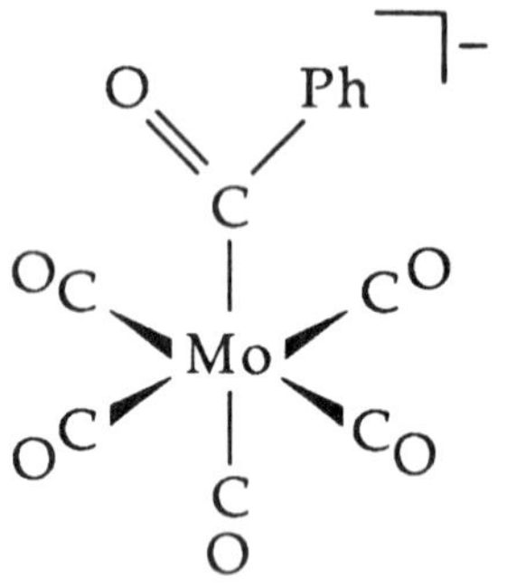

$[Mo(CO)_5(COPh)]^-$ 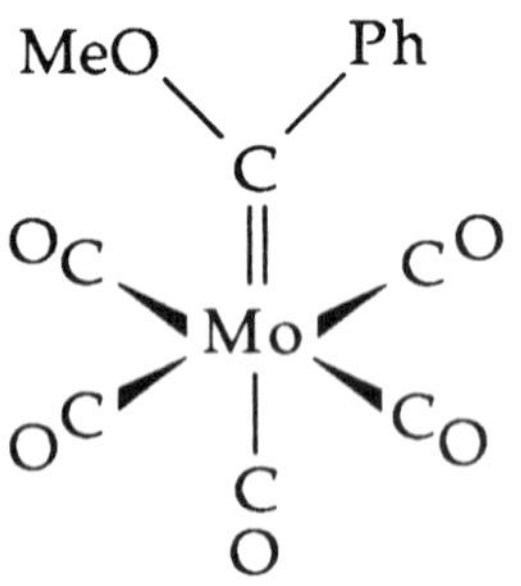              $Mo(CO)_5(C(OMe)Ph)$

**16.17 Explain the difference in IR spectra of the following:** (a) **Mo(PF₃)₃(CO)₃ vs. Mo(PMe₃)₃(CO)₃.** The two CO bands of the trimethylphosphine complex are 100 cm⁻¹ or more lower in frequency than the corresponding bands of the trifluorophosphine complex. This is because $PMe_3$ is primarily a σ-donor ligand while $PF_3$ is primarily a π-acid ligand ($PF_3$ is the ligand that most resembles CO electronically). The CO ligands in $Mo(PF_3)_3(CO)_3$ have to compete with the $PF_3$ ligands for electron density from the Mo atom for back donation. Therefore, less electron density is transferred from the Mo atom to the CO ligands in $Mo(PF_3)_3(CO)_3$ than in $Mo(PMe_3)_3(CO)_3$. This makes the Mo–C bonds in $Mo(PF_3)_3(CO)_3$ weaker than those in $Mo(PMe_3)_3(CO)_3$, but it also makes the C–O bonds in $Mo(PF_3)_3(CO)_3$ stronger than those in $Mo(PMe_3)_3(CO)_3$, and stronger C–O bonds will have higher CO stretching frequencies.

**(b)   $MnCp(CO)_3$ vs. $MnCp^*(CO)_3$.** The two CO bands of the $Cp^*$ complex are slightly lower in frequency than the corresponding bands of the Cp complex. Recall that the $Cp^*$ ligand is $\eta^5$-$C_5Me_5$. It is a stronger donor ligand than Cp, due to the inductive effect of the five methyl groups. Therefore, the Mn atom in the $Cp^*$ complex has a greater electron density than in the Cp complex, and hence there is a greater degree of back donation to the  CO ligands in the $Cp^*$ complex than in the Cp complex. As explained in part **(a)**, above, more back donation leads to lower stretching frequencies.

**16.18  Stability and reactivity of Cp complexes. (a)  Which is more stable, $RhCp_2$ or $RuCp_2$?** Refer to **Figure 16.12**, the MO diagram for metallocenes. The number of valence electrons for the two complexes differs by one; $RhCp_2$ has 19 electrons while $RuCp_2$ has 18 electrons. The MOs of ruthenocene are filled up to the very weakly bonding $a_1'$ orbital (see **Figure 16.13**). Its electron configuration, starting with the $e_2''$ orbitals, is $(e_2'')^4(a_1')^2$. Rhodocene, with its extra electron, has a $(e_2'')^4(a_1')^2(e_1'')^1$ configuration. Since the $e_1''$ orbitals are antibonding, the Rh–C bonds in rhodocene will be longer than the Ru–C bonds in ruthenocene (refer to **Table 16.10** and compare the Fe–C and Co–C bond distances in $FeCp_2$ and $CoCp_2$, which are 2.06 Å and 2.12 Å, respectively, and which are isoelectronic with $RhCp_2$ and $Ru(Cp_2)$). Therefore, ruthenocene is the more stable of these two complexes.

**(b)   A workable reaction for the conversion of $(\eta^5$-$C_5H_5)_2Fe$ to $(\eta^5$-$C_5H_5)(\eta^5$-$C_5H_4COCH_3)Fe$.** The Cp rings in cyclopentadienyl complexes behave like simple aromatic compounds such as benzene, and so are subject to typical reactions of aromatic compounds such as Freidel-Crafts alkylation and acylation.    If we treat ferrocene with acetyl chloride and some aluminum(III) chloride as a catalyst, we will obtain the desired compound:

$$(\eta^5\text{-}C_5H_5)_2Fe + CH_3COCl \xrightarrow{AlCl_3} (\eta^5\text{-}C_5H_5)(\eta^5\text{-}C_5H_4COCH_3)Fe + HCl$$

**16.19  A workable set of reactions for the conversion of $(\eta^5\text{-}C_5H_5)_2Fe$ to $(\eta^5\text{-}C_5H_5)(\eta^5\text{-}C_5H_4COOH)Fe$.**   In this case, we cannot use Friedel-Crafts chemistry to prepare the desired compound.  If we tried to prepare a –COOH substituent on one of the Cp rings by oxidation of a methyl group added to the Cp ring by Friedel-Crafts methylation, we would wind up using such a strong oxidant that it would destroy the complex.  For example, toluene can be oxidized to benzoic acid, but only using very strong oxidizing agents such as permanganate:

$$C_6H_5CH_3 \xrightarrow{MnO_4^-} C_6H_5COOH$$

Instead, we must find a milder way to add a carboxylic acid functionality to a Cp ring.  At the end of **Section 16.8** is a discussion about the lithiation of ferrocene, along with the following reaction:

$$(\eta^5\text{-}C_5H_5)_2Fe + BuLi \rightarrow (\eta^5\text{-}C_5H_5)(\eta^5\text{-}C_5H_4Li)Fe + BuH$$

If we treat lithiated ferrocene with $CO_2$ and then protonate that product, we will have succeeded in preparing $(\eta^5\text{-}C_5H_5)(\eta^5\text{-}C_5H_4COOH)Fe$:

$$(\eta^5\text{-}C_5H_5)(\eta^5\text{-}C_5H_4Li)Fe + CO_2 \rightarrow Li[(\eta^5\text{-}C_5H_5)(\eta^5\text{-}C_5H_4CO_2)Fe]$$

$$Li[(\eta^5\text{-}C_5H_5)(\eta^5\text{-}C_5H_4CO_2)Fe] + HCl \rightarrow (\eta^5\text{-}C_5H_5)(\eta^5\text{-}C_5H_4COOH)Fe + LiCl$$

**16.20  The $a_1'$ molecular orbital of a $D_{5h}$ Cp$_2$M complex.**   The symmetry-adapted orbitals of the two eclipsed $C_5H_5$ rings in a metallocene are shown in **Figure 16.13**.  As shown in that figure and in **Appendix 6**, the $d_{z^2}$ orbital on the metal has $a_1'$ symmetry and so can form MOs with this symmetry-adapted orbital.  **Appendix 6** shows that the $s$ orbital on the metal also has $a_1'$

symmetry, so it can form MOs with the $d_{z^2}$ orbital as well as the symmetry-adapted orbital. Therefore, three $a_1'$ MOs will be formed, since three orbitals of $a_1'$ symmetry are available on the metal and the ligands. **Figure 16.12** shows the energies of the three $a_1'$ MOs; one is the most stable orbital of a metallocene involving metal-ligand bonding, one is relatively nonbonding (it is the HOMO in FeCp$_2$), and one is a relatively high-lying antibonding orbital.

**16.21  Provide an explanation for the different sites of protonation of NiCp$_2$ and FeCp$_2$.**  Protonation of FeCp$_2$ at iron does not change its number of valence electrons:  Both FeCp$_2$ and [HFeCp$_2$]$^+$ are 18-electron complexes:

| FeCp$_2$ | | [HFeCp$_2$]$^+$ | |
|---|---|---|---|
| Fe | 8 valence $e^-$ | Fe | 8 valence $e^-$ |
| 2Cp | 10 $e^-$ | 2Cp | 10 $e^-$ |
| | ——— | H | 1 $e^-$ |
| | | +1 charge | $-1\ e^-$ |
| | | | ——— |
| Total | 18 $e^-$ | | 18 $e^-$ |

By the same token, since NiCp$_2$ is a 20-electron complex, the hypothetical metal-protonated species [HNiCp]$^+$ would also be a 20-electron complex. On the other hand, protonation of NiCp$_2$ at a Cp C atom produces the 18-electron complex [NiCp($\eta^4$-C$_5$H$_6$)]$^+$. Therefore, the reason that the Ni complex is protonated at a C atom is that a more stable (i.e., 18-electron) product is formed:

[NiCp($\eta^4$-C$_5$H$_6$)]$^+$

| | |
|---|---|
| Ni | 10 valence $e^-$ |
| Cp | 5 $e^-$ |
| $\eta^4$-C$_5$H$_6$ | 4 $e^-$ |
| +1 charge | $-1\ e^-$ |
| | ——— |
| Total | 18 $e^-$ |

**16.22 Write a plausible mechanism for the following reactions:**
**(a)** $[Mn(CO)_5(CF_2)]^+ + H_2O \rightarrow [Mn(CO)_6]^+ + 2HF$. The $=CF_2$ carbene ligand is similar electronically to a Fischer carbene (see **Section 16.7**). The electronegative F atoms render the C atom subject to nucleophilic attack, in this case by a water molecule. The $[(CO)_5Mn(CF_2(OH_2))]^+$ complex can then eliminate two equivalents of HF, as shown in the mechanism below:

$$[(CO)_5Mn=CF_2]^+ + H_2O \longrightarrow [(CO)_5Mn-C-O]^+ \quad \xrightarrow{-HF}$$

$$[(CO)_5Mn-C=O]^+ \quad \xrightarrow{-HF} \quad [(CO)_5Mn-C\equiv O]^+ \quad = [Mn(CO)_6]^+$$

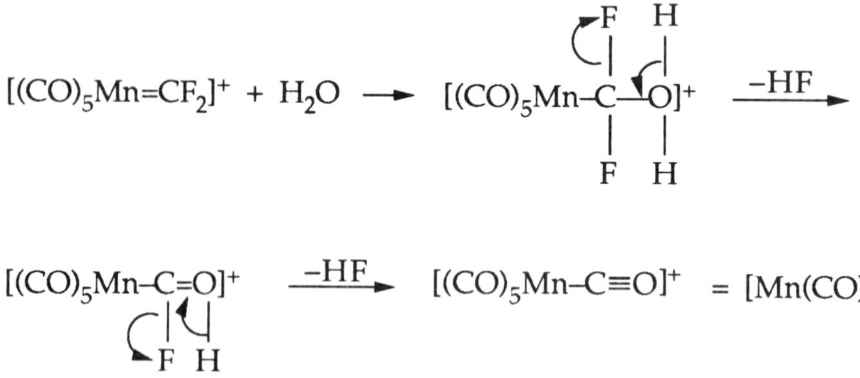

**(b)** $Rh(C_2H_5)(PR_3)_2(CO) \rightarrow RhH(PR_3)_2(CO) + C_2H_4$. This is an example of a β-hydrogen elimination reaction, discussed at the beginning of **Section 16.7**. This reaction is believed to proceed through a cyclic intermediate involving a *3c,2e* M–H–C interaction, as shown in the mechanism below:

$$(PR_3)_2(CO)Rh-C_2H_5 \longrightarrow$$

$$(PR_3)_2(CO)Rh-CH_2=CH_2 \longrightarrow RhH(PR_3)_2(CO) + C_2H_4$$

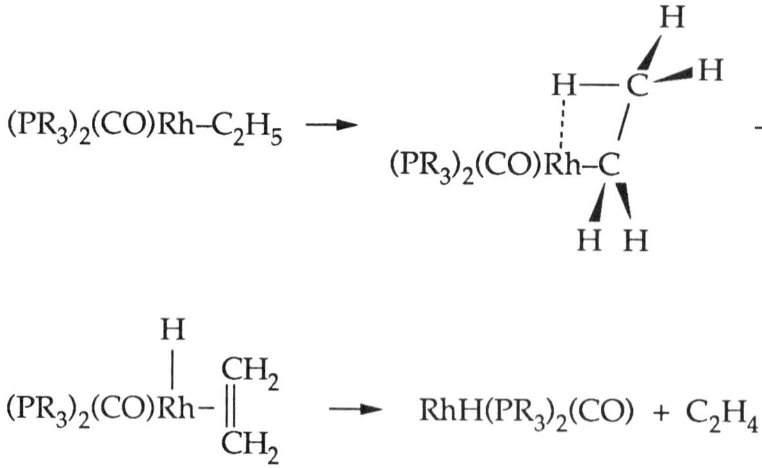

**16.23 Contrast Groups 6-8 with Groups 3 and 4 with respect to: (a) Stability of complexes of the $\pi$-$C_5H_5$ ligand.** The cyclopentadienyl ligand forms very stable complexes with metals from Groups 6-8. It can also form stable complexes with metals from Groups 3 and 4 (e.g. $Cp_2TiCl_2$), but in many cases an unwanted side reaction is activation of one of the C–H bonds of the ligand. An example of this behavior was discussed in **Section 16.9**: attempts to prepare $TiCp_2$ led instead to the dimer shown in **Structural Drawing 50**.

**(b) Hydridic or protonic character of M–H bonds.** Group 3 and Group 4 hydride complexes are truly hydridic. That is, the M–H bond in these complexes is polarized as follows:

$$M^{\delta+}\!-\!H^{\delta-}$$

In contrast, hydride complexes of Groups 6–8 have M–H bonds that are polarized in the opposite sense, and are not hydridic at all. In fact, they behave like weak acids. The difference helps to explain the fact that hydride complexes of the early *d*-block metals react with Brønsted acids as weak as alcohols, while hydride complexes of metals from Groups 6-8 do not:

$$HTiCp_2(CH_3) + ROH \rightarrow (RO)TiCp_2(CH_3) + H_2$$

$$HMn(CO)_5 + ROH \rightarrow \text{no reaction}$$

**(c) Adherence to the 18-electron rule.** Organometallic hydride complexes of the early *d*-block metals rarely adhere to the 18-electron rule. A good example is the titanium complex in part **(b)**, above: $HTiCp_2(CH_3)$ has 16 valence electrons. Another example is $HSc(\eta^5\text{-}C_5Me_5)_2$, which is also a 16-electron complex. In contrast, all stable hydride complexes of metals from Groups 6-8 have 18 valence electrons.

**16.24 Metal-metal bond orders for $Mo_2(py)_2(OAc)_4$ and $Mn_2(CO)_{10}$.** The usual energy ordering for bonding and antibonding orbitals in *d*-block dimetal complexes is shown on the left side of the figure below. In terms of increasing energy, the order is $\sigma < \pi < \delta < \delta^* < \pi^* < \sigma^*$. The complex $Mn_2(CO)_{10}$ contains two $d^7$ $Mn^0$ metal centers, so there are 14 electrons to place in the stack of M–M orbitals. Only $\sigma^*$ is left empty, so the bond order is

1. The complex $Mo_2(py)_2(OAc)_4$ contains two $d^4$ $Mo^{2+}$ metal conters, so there are 8 electrons to place in the stack. All of the bonding orbitals are filled, and all of the antibonding orbitals are empty, so the bond order is 4.

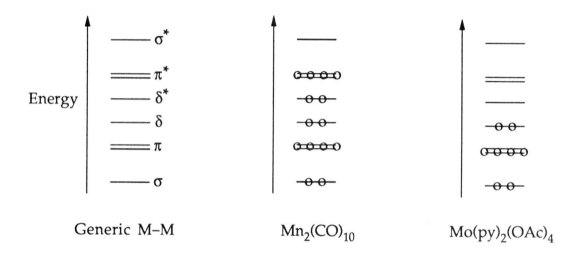

16.25 **Metal clusters and cluster valence electron (CVE) counts. (a) What CVE count is characteristic of octahedral and trigonal prismatic clusters?** According to **Table 16.12**, an octahedral $M_6$ will have 86 cluster valence electrons and a trigonal prismatic $M_6$ cluster will have 90.

   **(b) Can these CVE values be derived from the 18-electron rule?** No. As discussed in **Section 16.10**, the bonding in smaller clusters can be explained in terms of local M–M and M–L electron pair bonding and the 18-electron rule, but the bonding in octahedral $M_6$ and larger clusters cannot.

   **(c) Determine the probable geometry of $[Fe_6(C)(CO)_{16}]^{2-}$ and $Co_6(C)(CO)_{16}$.** The iron complex has 86 CVEs, while the cobalt complex has 90 (see the calculations below). Therefore, the iron complex probably contains an octahedral $Fe_6$ array, while the cobalt complex probably contains a trigonal prismatic $Co_6$ array.

| 6Fe | $6 \times 8e^- = 48\ e^-$ | 6Co | $6 \times 9\ e^- = 54\ e^-$ |
|---|---|---|---|
| C | $4\ e^-$ | C | $4\ e^-$ |
| 16CO | $16 \times 2e^- = 32\ e^-$ | 16CO | $16 \times 2e^- = 32\ e^-$ |
| −2 charge | $2\ e^-$ | | |
| Totals | $86\ e^-$ | | $90\ e^-$ |

**16.26  Choose the groups that might replace the following:  (a)  The CH group in $Co_3(CO)_9CH$:  $OCH_3$, $N(CH_3)_2$, or $SiCH_3$.**  Isolobal groups have the same number and shape valence orbitals *and* the same number of electrons in those orbitals.  The CH group has three $sp^3$ hybrid orbitals that each contain a single electron.  The $SiCH_3$ group has three similar orbitals similarly occupied, so it is isolobal with CH and would probably replace it in $Co_3(CO)_9CH$ to form $Co_3(CO)_9SiCH_3$.  In contrast, the $OCH_3$ and $N(CH_3)_2$ groups are not isolobal with CH.  Instead, they have, respectively, three $sp^3$ orbitals that each contain a pair of electrons and two $sp^3$ orbitals that each contain a pair of electrons.

**(b)  One of the $Mn(CO)_5$ groups in $(OC)_5MnMn(CO)_5$:  I, $CH_2$, or $CCH_3$.**  The $Mn(CO)_5$ group has a single σ type orbital that contains a single electron.  An iodine atom is isolobal with it, since it also has a singly occupied σ orbital.  Therefore, we can expect the compound $Mn(CO)_5I$ to be reasonably stable.  In contrast, the $CH_2$ and $CCH_3$ are not isolobal with $Mn(CO)_5$.  The $CH_2$ group has either a doubly occupied σ orbital and an empty *p* orbital or a singly occupied σ orbital and a singly occupied *p* orbital.  The $CCH_3$ group has three singly occupied σ orbitals (note that it is isolobal with $SiCH_3$.

$Mn(CO)_5I$ is
a stable complex

**16.27  Which cluster undergoes faster associative exchange with $^{13}CO$, $Co_4(CO)_{12}$ or $Ir_4(CO)_{12}$?**  If the rate determining step in the substitution is cleavage of one of the metal-metal bonds in the cluster, the cobalt complex will exhibit faster exchange. This is because metal-metal bond strengths *increase* down a group in the *d*-block. Therefore, Co-Co bonds are weaker than Ir-Ir bonds, all other things like geometry and ligand types kept the same.

# CHAPTER 17

# CATALYSIS

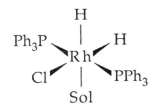

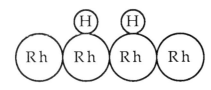

There are many chemical parallels between homogeneous and heterogeneous catalysis, despite the different terminology used to describe them and the different physical methods used to study them. For example, soluble rhodium complexes and rhodium metal both react with hydrogen. In the case of the soluble complex, we call the reaction an oxidative addition; in the case of rhodium metal, we call the reaction a chemisorption. In both cases, however, reactive Rh–H bonds are formed.

---

## SOLUTIONS TO IN-CHAPTER EXERCISES

**17.1 Oxidative additions and reductive eliminations.** In **Figure 17.1**, the oxidative addition is the reaction in which the *four*-coordinate complex $RhClL_2(Sol)$ reacts with $H_2$ to form the *six*-coordinate complex $RhClH_2L_2(Sol)$. The reaction is called an addition because of the increase in coordination number. The balanced equation is (Sol = a solvent molecule):

$$RhClL_2(Sol) + H_2 \rightarrow RhClH_2L_2(Sol)$$

Since chloride and hydride ligands are normally considered as anionic ligands ($Cl^-$ and $H^-$), the oxidation numbers for rhodium are +1 for $RhClL_2(Sol)$ and +3 for $RhClH_2L_2(Sol)$. The reductive elimination in the figure is the reaction in which the six-coordinate complex $RhCl(C_2H_4R)HL_2(Sol)$ eliminates alkane and forms the four-coordinate complex $RhClL_2(Sol)$. The balanced equation is:

$$RhCl(C_2H_4R)HL_2(Sol) \rightarrow C_2H_5R + RhClL_2(Sol)$$

The alkyl ligand is also normally considered to be anionic ($C_2H_4R^-$), so the oxidation numbers for the rhodium complexes in this reaction are +3 for $RhCl(C_2H_4R)HL_2(Sol)$ and +1 for $RhClL_2(Sol)$.

**17.2 The effect of added phosphine on the catalytic activity of $RhH(CO)(PPh_3)_3$.** As implied in the preceeding section, $RhH(CO)(PPh_3)_3$ is an 18-electron complex that must lose a phosphine ligand before it can enter the catalytic cycle:

$$RhH(CO)(PPh_3)_3 \rightleftharpoons RhH(CO)(PPh_3)_2 + PPh_3$$
$$\text{18 valence } e^- \qquad \text{16 valence } e^- \qquad 2e^- \text{ donor}$$

The coordinatively unsaturated $RhH(CO)(PPh_3)_2$ complex can add the alkene that is to be hydroformylated. Added phosphine will shift the above equilibrium to the left, resulting in a lower concentration of the catalytically active 16-electron complex. Thus, we can predict that the rate of hydroformylation will be decreased by added phosphine.

**17.3 γ-Alumina heated to 900 °C, cooled, and exposed to pyridine vapor.** We saw in **Section 6.9** that heating alumina to 900 °C results in complete dehydroxylation (see **Figure 6.5**). Therefore, only Lewis acid sites are present in γ-alumina heated to 900 °C. The IR spectrum of a sample of dehydroxylated γ-alumina exposed to pyridine will exhibit bands near 1465 cm$^{-1}$. It will not exhibit bands near 1540 cm$^{-1}$, since these are due to pyridine that is hydrogen bonded to the surface, and no surface OH groups are present in dehydroxylated alumina.

**17.4    Would a pure silica analog of ZSM-5 be an active catalyst for benzene alkylation?**    A pure silica analog of ZSM-5 would contain Si–OH groups, which are only moderate Brønsted acids, and not the strongly acidic Al–OH$_2$ groups found in aluminosilicates (see **Structural Drawing 17.6** and **Section 6.9**).    Only a very strong Brønsted acid such as an Al–OH$_2$ group in an aluminosilicate can protonate an alkene to form the carbocations that are necessary intermediates in benzene alkylation.    Therefore, a pure silica analog of ZSM-5 would not be an active catalyst for benzene alkylation.

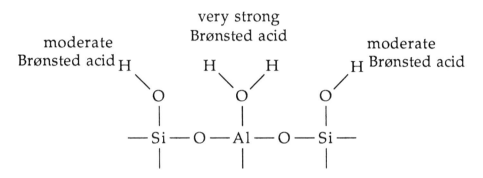

---

## SOLUTIONS TO END-OF-CHAPTER EXERCISES

**17.1    Which of the following constitute catalysis?    (a)    H$_2$ and C$_2$H$_4$ in contact with Pt.**    This is an example of genuine catalysis.    The formation of ethane (H$_2$ + C$_2$H$_4$ → C$_2$H$_6$) has a very high activation barrier.    The presence of platinum causes the reaction to proceed at a useful rate.    Furthermore, the platinum can be recovered unchanged after many turnovers, so it fits the two criteria of a catalyst, a substance that increases the rate of a reaction but is not itself consumed.

(b)    **H$_2$ plus O$_2$ plus an electrical arc.**    This gas mixture will be completely converted into H$_2$O once the electrical arc is struck. Nevertheless, it does not constitute an example of catalysis.    The arc provides activation energy to initiate the reaction.    Once started, the heat liberated by the reaction provides activation energy to sustain the reaction.    The activation energy of the reaction has *not* been lowered by an added substance, so catalysis has not occured.

(c) **The production of Li$_3$N and its reaction with H$_2$O.** This too is not an example of catalysis. Lithium and nitrogen are both consumed in the formation of Li$_3$N, which occurs at an appreciable rate even at room temperature. Water and Li$_3$N react rapidly to produce NH$_3$ and LiOH at room temperature. As in the formation of Li$_3$N, both substances are consumed. Since both reactions have intrinsically low activation energies, a catalyst is not necessary.

**17.2    Define the following terms: (a) Turnover frequency.** This is defined differently for homogeneous and heterogeneous catalysis. In both cases, however, it is really the same thing, the *amount* of product formed per unit time per unit *amount* of catalyst. In homogeneous catalysis, the turnover frequency is the rate of formation of product, given in mol L$^{-1}$ s$^{-1}$, divided by the concentration of catalyst used, in mol L$^{-1}$. This gives the turnover frequency in units of s$^{-1}$. In heterogeneous catalysis, the turnover frequency is typically the amount of product formed per unit time, given in mol s$^{-1}$, divided by the number of moles of catalyst present. In this case, the turnover frequency also has units of s$^{-1}$. Since one mole of a finely divided heterogeneous catalyst is more active than one mole of the same catalyst with a small surface area, the turnover frequency is sometimes expressed as the amount of product formed per unit time divided by the surface area of the catalyst. This gives the turnover frequency in units of mol s$^{-1}$ cm$^{-2}$. Often one finds the turnover frequency for commercial heterogeneous catalysts expressed in rate per gram of catalyst.

(b) **Selectivity.** This is a measure of how much of the desired product is formed relative to undesired byproducts. Unlike enzymes (see **Chapter 19**), man-made catalysts rarely are 100% selective. The catalytic chemist usually has to deal with the often difficult problem of separating the various products. The separations, by distillation, fractional crystallization, or chromatography, are generally expensive and are always time consuming. Furthermore, the byproducts represent a waste of raw materials. Recall that an expensive rhodium hydroformylation catalyst is sometimes used industrially instead of a relatively inexpensive cobalt catalyst because the rhodium catalyst is more selective (see **Section 17.4**).

(c) **Catalyst.** The first line of **Chapter 17** reads "A catalyst is a substance that increases the rate of a reaction but is not itself consumed." This does not mean that the "catalyst" that is added to the reaction mixture is left unchanged during the course of the reaction. Frequently, the substance that is added is a *catalyst precursor* that is transformed under the reaction conditions into the active catalytic species.

(d) **Catalytic cycle.** This is a sequence of chemical reactions, each involving the catalyst, that transform the reactants into products. It is called a cycle because the actual catalytic species involved in the first step is regenerated during the last step. Note that the concept of a "first" and "last" step may lose its meaning once the cycle is started.

(e) **Catalyst support.** In cases where a heterogeneous catalyst does not remain a finely divided pure substance with a large surface area under the reaction conditions, it must be dispersed on a support material, which is generally a ceramic like $\gamma$-alumina or silica gel. In some cases, the support is relatively inert and only serves to maintain the integrity of the small catalyst particles. In other cases, the support interacts strongly with the catalyst and may affect the rate and selectivity of the reaction.

**17.3   Classify the following as homogeneous or heterogeneous catalysis:** (a) **The increased rate of $SO_2$ oxidation in the presence of NO.** The balanced equation for $SO_2$ oxidation is:

$$2SO_2\,(g) \;+\; O_2\,(g) \;\rightarrow\; 2SO_3\,(g)$$

All of these substances, as well as the catalyst, NO, are gases. Since they are all present in the same phase, this is an example of homogeneous catalysis (see **Section 17.1**).

(b) **The hydrogenation of oil using a finely divided Ni catalyst.** In this case, the balanced equation is:

$$RHC=CHR' + H_2 \rightarrow RH_2C–CH_2R'$$

The reactants and the products are all present in the liquid phase (the hydrogen is dissolved in the liquid oil), but the catalyst is a solid. Therefore, this is an example of heterogeneous catalysis.

(c) **The conversion of D-glucose to a D,L mixture by HCl.** The catalyst for the racemization of D-glucose is HCl (really $H_3O^+$), which is present in the same aqueous phase as the D-glucose. Therefore, since the substrate and the catalyst are both in the same phase, this is homogeneous catalysis.

D–Glucose                                                        L–Glucose

**17.4    Which of the following processes would be worth investigating?**
(a) **The splitting of $H_2O$ into $H_2$ and $O_2$.** A catalyst does not affect the free energy ($\Delta G$) of a reaction, only the activation free energy ($\Delta G^{\ddagger}$). A thermodynamically unfavorable reaction (one with a positive $\Delta G$) will not result in a useful amount of products unless energy is added in the form of light or electric current is added to the reaction mixture. For example, if $\Delta G \approx$ +20 kJ mol$^{-1}$, $K_{eq} \approx 10^{-3}$. Therefore, it would not be a worthwhile endeavor to try to develop a catalyst to split water into hydrogen and oxygen because water is an exoergic compound (it is stable with respect to its constituent elements; see **Table 9.5**).

$$H_2O \; (l) \; \rightarrow \; H_2 \; (g) \; + \; O_2 \; (g) \qquad \Delta G° = 237 \; kJ \; mol^{-1}$$

(b) **The decomposition of $CO_2$ into C and $O_2$.** As in part **(a)**, we would be trying to catalyze the decomposition of a very stable compound into its constituent elements. It would be a waste of time.

$$CO_2 \; (g) \; \rightarrow \; C \; (s) \; + \; O_2 \; (g) \qquad \Delta G° = 394 \; kJ \; mol^{-1}$$

(c) **The combination of $N_2$ with $H_2$ to produce $NH_3$.** This would be a very worthwhile reaction to try to catalyze efficiently at 80 °C. Ammonia is a stable compound with respect to nitrogen and hydrogen. Furthermore, it is a compound that is important in commerce, since it is used in many types of fertilizers. $\Delta G°$ is only $-16.5 \; kJ \; mol^{-1}$ for the reaction $N_2 + 3H_2 \rightarrow 2NH_3$ (see **Table 9.5**). The high temperatures usually required for ammonia synthesis (~400 °C) make $\Delta G$ less negative and result in a smaller yield. An efficient *low temperature* synthesis of ammonia from nitrogen and hydrogen that could be carried out on a large scale would probably make you and your industrialist rich.

(d) **The hydrogenation of double bonds in vegetable oil.** The reaction $RHC=CHR + H_2 \rightarrow RH_2C–CH_2R$ has a negative $\Delta G°$ (see **Figure 17.1** and **Section 17.4**). Therefore, the hydrogenation of vegetable oil would be a candidate for catalyst development. However, there are many homogeneous and heterogeneous catalysts for olefin hydrogenation, including those that operate at low temperatures (i.e., ~80 °C). For this reason, the process can be readily set up with existing technology.

**17.5    Explain why the addition of $PPh_3$ to $RhCl(PPh_3)_3$ reduces the hydrogenation turnover frequency.** The catalytic cycle for homogeneous hydrogenation of alkenes by Wilkinson's catalyst, $RhCl(PPh_3)_3$, is shown in **Figure 17.3**. Let us focus on the dominant path. The catalytic species that enters the cycle is not $RhCl(PPh_3)_3$ but $RhCl(PPh_3)_2(Sol)$ (Sol = a solvent molecule), formed by the following equilibrium:

$$RhCl(PPh_3)_3 \; + \; Sol \; \rightleftharpoons \; RhCl(PPh_3)_2(Sol) \; + \; PPh_3$$

Therefore, added PPh3 will shift this equilibrium to the left, resulting in a lower concentration of the active catalytic species. Note that this is the only reaction that added PPh3 can affect, since no other step involves free PPh3.

**17.6   Explain the trend in rates of H2 absorption by various olefins catalyzed by RhCl(PPh3)3.** The data show that hydrogenation is faster for hexene than for *cis*-4-methyl-2-pentene (a factor of 2910/990 = 3). It is also faster for cyclohexene than for 1-methylcyclohexene (a factor of 3160/60 = 53). In both cases, the alkene that is hydrogenated more slowly has a greater degree of substitution and so is sterically more demanding:   hexene is a monosubstituted alkene while *cis*-4-methyl-2-pentene is a disubstituted alkene; cyclohexene is a disubstituted alkene while 1-methylcyclohexene is a trisubstituted alkene.   According to the catalytic cycle for hydrogenation shown in **Figure 17.3**, different alkenes could affect the equilibrium (b) ⇌ (c) or the reaction (c) → (d). A sterically more demanding alkene could result in (i) a smaller equilibrium concentration of (c) (RhClH$_2$L$_2$(alkene)), or (ii) a slower rate of conversion of (c) to (d) (RhClHL$_2$(alkyl)(Sol)).   Since the reaction (c) → (d) is the rate determining step for hydrogenation by RhCl(PPh3)3, either effect would lower the rate of hydrogenation.

2910 L mol$^{-1}$ s$^{-1}$     990 L mol$^{-1}$ s$^{-1}$     3160 L mol$^{-1}$ s$^{-1}$   60 L mol$^{-1}$ s$^{-1}$

**17.7   Hydroformylation catalysis with and without added P(*n*-Bu)3.** Since compound (c) is observed under catalytic conditions, its formation must be faster than its transformation into products. If this were not true, a spectroscopically observable amount of it would not build up. Therefore, the transformation of (c) into CoH(CO)4 must be the rate determining step in the absence of added P(*n*-Bu)3.   In the presence of added P(*n*-Bu)3, neither compound (c) nor its phosphine-substituted equivalent Co(C(=O)C4H9)-

$(CO)_3(P(n\text{-}Bu)_3)$ is observed, requiring that their formation must be slower than their transformation into products. Thus, in the presence of added $P(n\text{-}Bu)_3$, the formation of either (a), (c), or their phosphine-substituted equivalents is the rate determining step.

**17.8 The two likely mechanisms for the Wacker process.** The two β-hydroxyethyl complexes that result from attack by free $OH^-$ or coordinated $OH^-$ on coordinated *trans*-1,2-$C_2H_2D_2$ are shown below. In the case of attack by coordinated $OH^-$, two views of the same product are shown, differing by a free rotation of 180° about the C–C bond. You can see that the two products are stereochemical isomers, *erythro* and *threo*. If only the *erythro* product were formed, for example, you could safely infer that the reaction proceeded by the attack on the olefin by free $OH^-$.

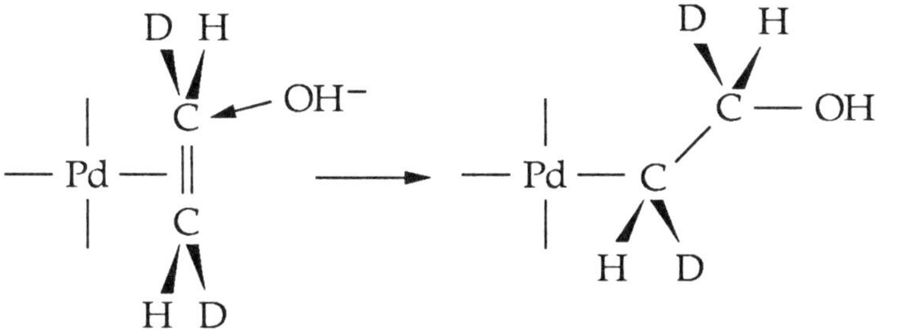


*erythro* isomer

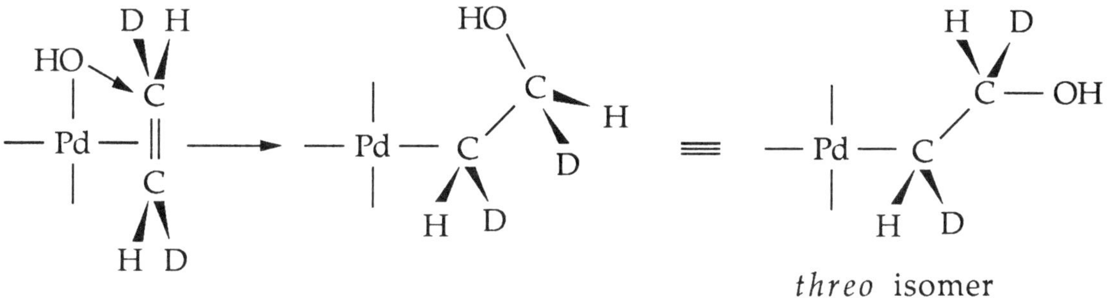


*threo* isomer

**17.9 A plausible mechanism for the catalysis of formic acid decomposition by $IrClH_2(PPh_3)_3$.** Formic acid can be decomposed by the following sequence of reactions: oxidative addition of the O–H bond, β-hydride elimination, reductive elimination of $H_2$, and $CO_2$ dissociation, as shown below for the generic *coordinatively unsaturated* metal complex $M^0$ (the reductive elimination of $H_2$ and $CO_2$ dissociation are combined into a single final step).

The iridium complex $IrClH_2(PPh_3)_3$ contains Ir(III), is six-coordinate, and is has 18 valence electrons. As such, it cannot coordinate to formic acid without prior dissociation of a phosphine ligand, and it cannot undergo oxidative addition of formic acid without prior reductive elimination of two other anionic ligands, such as the two hydride ligands. A plausible mechanism for catalysis by $IrClH_2(PPh_3)_3$ is as follows:

$$IrClH_2(PPh_3)_3 \rightarrow IrClH_2(PPh_3)_2 + PPh_3$$

$$IrClH_2(PPh_3)_2 + HCOOH \rightarrow IrClH_2(PPh_3)_2(HCOOH)$$

$$IrClH_2(PPh_3)_2(HCOOH) \rightarrow IrCl(PPh_3)_2(HCOOH) + H_2$$

$$IrCl(PPh_3)_2(HCOOH) \rightarrow IrClH(HCO_2)(PPh_3)_2$$

$$IrClH(HCO_2)(PPh_3)_2 \rightarrow IrClH_2(CO_2)(PPh_3)_3$$

$$IrClH_2(CO_2)(PPh_3)_3 \rightarrow IrClH_2(PPh_3)_2 + CO_2$$

**17.10 Why is titanium ineffective whereas iron is effective as a catalyst for ammonia synthesis?** The problem with titanium is clearly stated in the exercise. Titanium reacts with $N_2$ to form a *stable* nitride. If the stability of a catalytic intermediate is too great, it will build up and will not be transformed into products. Refer to **Figure 17.2**, which shows schematic representations of the energetics in a catalytic cycle. The formation of titanium nitride would be represented by the deep trough (the dashed line). An efficient catalytic cycle must have a relatively smooth free energy vs. extent of reaction graph, free of deep troughs and high peaks.

**17.11 Aluminosilicate surface acidity: (a) Give an explanation for the enhancement of acidity by the presence of $Al^{3+}$ in a silica lattice.** The substitution of Al(III) for Si(IV) in a silica lattice would lead to a charge imbalance (there would be one too few positive charges per Al/Si substitution). To compensate for the charge imbalance, an extra $H^+$ ion per Al/Si substitution is associated with the aluminosilicate lattice, as shown in the diagram accompanying the solution to **In-Chapter Exercise 17.4**. The $Al–OH_2$ groups are much more acidic than Si–OH groups.

**(b) Name three other ions that might enhance the acidity of silica gel.** A number of other +3 cations would lead to the same charge imbalance produced by substituting Al(III) for Si(IV). These include Sc(III), Cr(III), Fe(III), and Ga(III).

**17.12 Indicate the difference between each of the following: (a) The Lewis acidity of γ-alumina that has been heated to 900 °C versus γ-alumina that has been heated to 100 °C.** As discussed in **Section 6.9**, γ-alumina that has not been heated to above 150 °C is completely hydroxylated. This means that all of the surface Al(III) ions are four-coordinate, and as such they will not exhibit appreciable Lewis acidity. On the other hand, γ-alumina is completely dehydroxylated by heating to 900 °C. The surface of dehydroxylated γ-alumina contains three-coordinate, strongly acidic Al(III) sites, as shown in **Figure 6.5**.

**(b) The Brønsted acidity of silica gel versus γ-alumina.** While aluminosilicates are strong Brønsted acids (see the solution to **End-of-**

**Chapter Exercise 17.11**), silica gel and γ-alumina are not. The Si–OH groups found on the surface of silica gel are moderate Brønsted acids (about as acidic as acetic acid — see **Section 6.9**). The Al–OH groups found on the surface of γ-alumina are weak Brønsted acids. They are weaker than Si–OH groups because of the lower charge of Al(III) relative to Si(IV). Recall that for aqua ions the charge on the metal was directly related to the Brønsted acidity of the ion (see **Section 5.5**, especially **Figure 5.4**). For example, $Fe(H_2O)_6^{3+}$ is six orders of magnitude more acidic than $Fe(H_2O)_6^{2+}$.

(c) **The structural regularity of the channels and voids in silica gel versus that for ZSM-5.** As can be seen from a schematic diagram of silica gel, such as the one shown in **Figure 17.10**, the structure of silica gel is quite irregular: channels and voids of a variety of sizes are evident. In contrast, the structure of ZSM-5, which is shown in **Figure 17.11**, has channels and voids of uniform size. ZSM-5 is a metastable, crystalline material with a regular structure. Silica gel is a metastable amorphous solid.

**17.13 Why is the platinum-rhodium in automobile catalytic converters dispersed on the surface of a ceramic rather than used in the form of thin foil?** As discussed in **Section 17.5**, special measures are required in heterogeneous catalysis to ensure that the reactants achieve contact with catalytic sites. For a given amount of catalyst, the greater the surface area the greater the number of catalytic sites. A thin foil of platinum-rhodium will not have as much surface area as an equal amount of small particles finely dispersed on the surface of a ceramic support. The diagram below shows this for a "foil" of catalyst that is 1000 times larger in two dimensions than in the third dimension (i.e., the thickness is 1; note that the diagram is not to scale). If the same amount of catalyst is broken into cubes that are 1 unit on a side, the surface area of the catalyst is increased by nearly a factor of three.

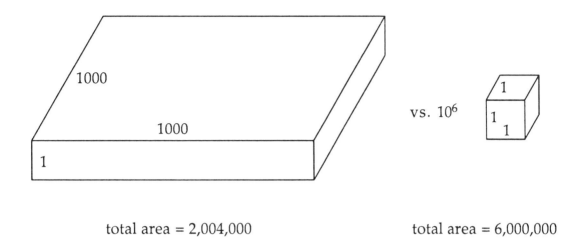

total area = 2,004,000                    total area = 6,000,000

**17.14 Describe the concept of shape-selective catalysts.** Certain catalysts such as zeolites have channels and voids of specific sizes and shapes (see **Section 11.11**). It has been found that the product distribution can be changed by switching from one zeolite to another with different size channels and voids. There are two explanations, as described in **Section 17.7**. The first is that products of certain sizes diffuse through the zeolite lattice more quickly than others. The remaining products, with longer residence times, can isomerize to products that diffuse away quickly. According to this view, the relative rate of diffusion of a product *once it is formed* is the key property. The currently favored view is that the sizes and shapes of the channels and voids allow only certain products to be formed in the first place.

**17.15 Devise a plausible mechanism to explain the deuteration of 3,3-dimethylpentane.** There are *two* observations that must be explained. The first is that ethyl groups are deuterated before methyl groups. The second observation is that a given ethyl group is completely deuterated before another one incorporates *any* deuterium. Let's consider the first observation first. The mechanism of deuterium exchange is probably related to the reverse of the last two reactions in **Figure 17.16**, which shows a schematic mechanism for the hydrogenation of an olefin by $D_2$. The steps necessary for deuterium substitution into an alkane are shown below, and include the dissociative chemisorption of an R–H bond, the dissociative chemisorption of $D_2$, and the dissociation of R–D. This can occur many times with the same alkane molecule to effect complete deuteration.

$$
\text{— Pt — Pt — Pt — Pt —} \;+\; \text{RH} \;\longrightarrow\;
\begin{array}{cc}
\text{H} & \text{R} \\
| & | \\
\end{array}
\text{— Pt — Pt — Pt — Pt —}
$$

$$
\begin{array}{cc}
\text{H} & \text{R} \\
| & | \\
\end{array}
\text{— Pt — Pt — Pt — Pt —} \;+\; \text{D}_2 \;\longrightarrow\;
\begin{array}{cccc}
\text{H} & \text{R} & \text{D} & \text{D} \\
| & | & | & | \\
\end{array}
\text{— Pt — Pt — Pt — Pt —}
$$

$$
\begin{array}{cccc}
\text{H} & \text{R} & \text{D} & \text{D} \\
| & | & | & | \\
\end{array}
\text{— Pt — Pt — Pt — Pt —} \;\longrightarrow\; \text{RD} \;+\;
\begin{array}{cc}
\text{H} & \text{D} \\
| & | \\
\end{array}
\text{— Pt — Pt — Pt — Pt —}
$$

Since the $-CH_2CH_3$ groups are deuterated before the $-CH_3$ groups, one possibility is that dissociative chemisorption of a C–H bond from a $-CH_2-$ group is faster than dissociative chemisorption of a C–H bond from a $-CH_3$ group. The second observation can be explained by invoking a mechanism for rapid deuterium exchange of the methyl group in the chemisorbed $-CHR(CH_3)$ group (R = $C(CH_3)_2(C_2H_5)$). The scheme below shows such a mechanism. It involves the succesive application of the equilibrium shown in **Figure 17.16**. If this equilibrium is maintained more rapidly than the dissociation of the alkane from the metal surface, the methyl group in question will be completely deuterated before dissociation takes place.

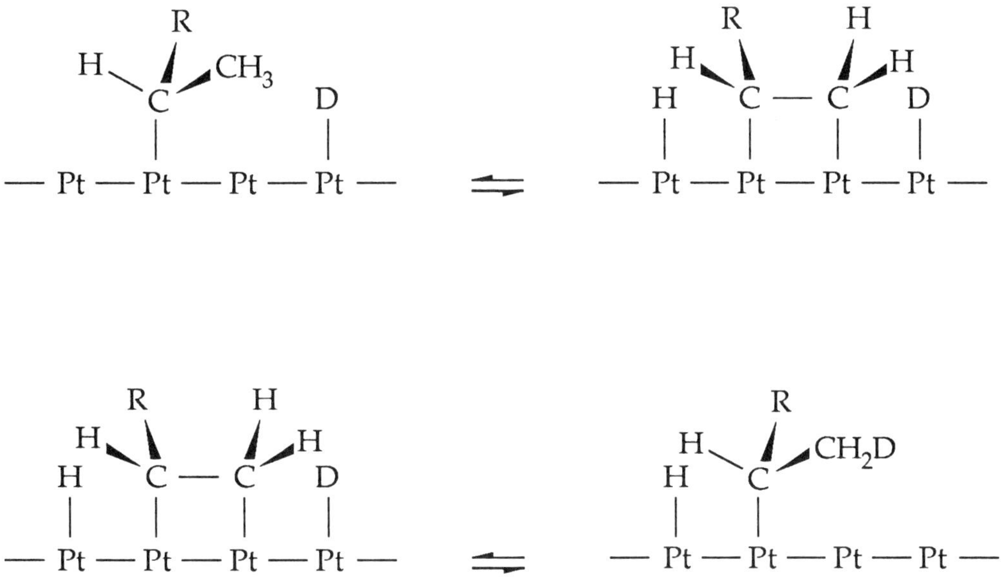

R
H\ /CH₃
 C        D
 |        |
— Pt — Pt — Pt — Pt —   ⇌

R        H
H\ \  / /H
H   C — C   D
 |   |   |   |
— Pt — Pt — Pt — Pt —

R        H
H\ \  / /H
H   C — C   D
 |   |   |   |
— Pt — Pt — Pt — Pt —   ⇌

R
H\ /CH₂D
H   C
 |   |
— Pt — Pt — Pt — Pt —

Another possibility, not shown in the scheme above, is that the terminal –CH₃ groups undergo more rapid dissociative chemisorption than the sterically more hindered internal –CH₃ groups.

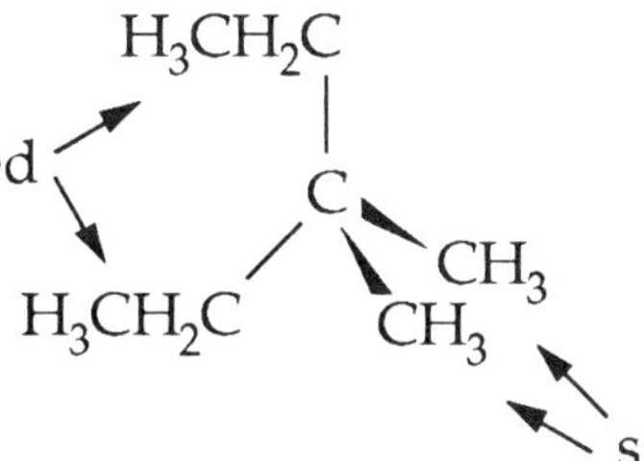

H₃CH₂C
        |
sterically less hindered     C
H₃CH₂C    CH₃
          CH₃
                sterically more hindered

**17.16 Why does CO decrease the effectiveness of Pt in catalyzing the reaction $2H^+ (aq) + 2e^- \rightarrow H_2 (g)$?** The reduction of hydrogen ions to $H_2$ probably involves the formation of surface hydride species similar to the ones shown in the figure on the opening page of this chapter. Dissociation of $H_2$ by reductive elimination would complete the catalytic cycle. According to **Table 17.5**, platinum not only has a strong tendency to chemisorb $H_2$, but it also has a strong tendency to chemisorb CO. If the surface of platinum is covered with CO, the number of catalytic sites available for H$^+$ reduction will be greatly diminished and the rate of H$_2$ production will decrease.

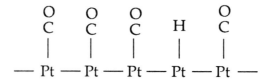

# CHAPTER 18

# STRUCTURES AND PROPERTIES OF SOLIDS

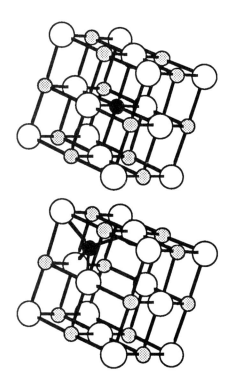

All solids contain defects, which influence properties such as mechanical strength and conductivity. The top figure shows the ideal structure of AgCl with no defects (the $Cl^-$ ions are the larger open spheres). All of the $Ag^+$ ions, including the highlighted black one, are in octahedral holes of $Cl^-$ ions. In a real sample of AgCl, some of the $Ag^+$ ions are displaced from octahedral sites to tetrahedral sites, as shown for the black $Ag^+$ ion in the bottom figure. This is an example of a Frenkel defect, which is a type of intrinsic point defect.

## SOLUTIONS TO IN-CHAPTER EXERCISES

**18.1   Does the measured density of VO indicate the presence of vacancies or interstitials?**   As in the **Example**, a measured density that is lower than the calculated density is indicative of vacancies (Schottky defects). This is because the volume of the crystal remains essentially constant, but some of the mass is missing. If the measured and calculated densities are nearly the same, the defects are probably interstitial (Frenkel defects). In the case of VO, the measured density is only about 91% of the calculated density. For the 1:1 ratio to be maintained, the vacancies must be on both V and O sites.

**18.2    Why does increased pressure reduce the conductivity of $K^+$ more than $Na^+$ in $\beta$-alumina?**    Increasing the pressure on a crystal compresses it, reducing the spacings between ions (for example, subjecting NaCl to a pressure of 24,000 atm reduces the Na–Cl distance from 2.82 Å to 2.75 Å). In a rigid lattice such as $\beta$-alumina, larger ions migrate slower than smaller ions with the same charge (as discussed in the **Example**). At higher pressures, with smaller conduction plane spacings, *all* ions will migrate more slowly than they do at atmospheric pressure. However, larger ions will be impeded to a greater extent by smaller spacings than will smaller ions. This is because the ratio (radius of migrating ion)/(conduction plane spacing) changes more, per unit change in conduction plane spacing, for a large ion than for a small ion. Therefore, increased pressure reduces the conductivity of $K^+$ more than $Na^+$ because $K^+$ is larger than $Na^+$.

**18.3    How many spins per formula unit of $NiFe_2O_4$ are expected at low temperatures?**    This compound exhibits an inverse spinel structure (see Table 18.3; $\lambda = 0.5$, so the structure can be denoted as Fe[NiFe]O₄). Half of the $Fe^{3+}$ ions are in tetrahedral sites and half are in octahedral sites, and all of the $Ni^{2+}$ ions are in octahedral sites. The $d$-orbital splitting diagrams for octahedral $Ni^{2+}$ and $Fe^{3+}$ are shown below.

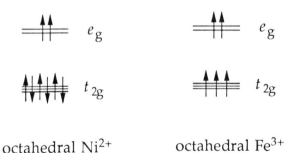

octahedral $Ni^{2+}$          octahedral $Fe^{3+}$

As stated in the **Example**, antiferromagnetic coupling between the tetrahedral and octahedral $Fe^{3+}$ ions would lead to cancellation of the five spins per ion at low temperatures. This would leave two spins per formula unit, because an octahedral $d^8$ ion such as $Ni^{2+}$ has two unpaired electrons in its $e_g$ orbitals.


**18.4   Should $Mo_4Ru_2Se_8$ exhibit metallic conduction or semiconduction?** Recall that a simple MO treatment of the $Mo_6S_8$ cluster indicates that it has the capacity for a total of 24 electrons (20 cluster valence electrons and 4 nonbonding electrons). Since $Mo_4Ru_2Se_8$ has 24 metal electrons (see the **Example**), the $d$ orbital band will be completely filled (this assumes that the frontier MOs for $Mo_4Ru_2Se_8$ are essentially the same as for the parent compound $Mo_6S_8$). For conduction to occur, an electron from the filled $d$ orbital band has to be promoted to the lowest energy empty band, which is separated from the $d$ orbital band by a nonzero bandgap. Thus, semiconduction will be observed for $Mo_4Ru_2Se_8$. Note that if it were possible to intercalate into $Mo_4Ru_2Se_8$ a guest that removed an electron or two per intercalant from the $d$ orbital band, the intercalated material would exhibit metallic conduction.


# SOLUTIONS TO END-OF-CHAPTER EXERCISES

**18.1   Describe the nature of Frenkel and Schottky defects.** These are both called intrinsic point defects. Intrinsic means that substances that contain these defects may be pure substances — they do not have to contain impurities. Point defects occur at single sites in the lattice and cannot be detected by X-ray crystallography or by electron microscopy, in contrast with extended defects. Frenkel and Schottky defects differ in the way the single sites are affected. A Frenkel defect involves an atom or ion moving from its usual site to an interstitial site (see the figure on the opening page of this chapter). Frenkel defects are difficult to verify experimentally — usually their presence is inferred from ionic conductivity data. A Schottky defect is a vacancy in an otherwise perfect lattice. At low vacancy concentrations, the displaced atoms can take up sites on the crystal surface. Since a substance with Schottky defects contains vacancies, the measured density is lower than the calculated density. The stoichiometry of substances that contain Frenkel and Schottky defects is not changed by the presence of the defects.


**18.2    Pick the compound most likely to have a high concentration of defects: (a)  NaCl or NiO.** To answer this question, the two things you want to consider are the openness of the structure (open structures provide larger sites that can accomodate interstitial atoms) and the possibility of variable oxidation states.  In this case the two compounds both have the rock salt structure (see **Table 12.9**).  Furthermore, inspection of **Table 4.2** should lead to the conclusion that ionic radii of $Na^+$ and $Ni^{2+}$ do not differ a great deal (the table contains the radii for $Na^+$ and $Ca^{2+}$ — $Ni^{2+}$ will be somewhat smaller than $Ca^{2+}$ due to imperfect shielding by $d$ electrons).  Therefore, the openness of the structure is not an issue.  Variable oxidation states are not possible for sodium, but they are for nickel.  NiO could contain some $Ni^{2+}$ ion vacancies, with two remaining $Ni^{2+}$ ions oxidized to $Ni^{3+}$ to compensate for each $Ni^{2+}$ vacancy (see **Section 18.4**).  Thus, you should expect NiO to contain more defects than NaCl.  The defects will be $Ni^{3+}$ ions occupying $Ni^{2+}$ sites and vacancies on the $Ni^{2+}$ sublattice.

(b)  **$CaF_2$ or $PbF_2$.** In this case too, the two compounds have the same structure (the fluorite structure).  However, unlike $Na^+$ vs. $Ni^{2+}$, the two cations in question here have very different ionic radii, and $Pb^{2+}$ is much more polarizable.  Therefore, you should expect $PbF_2$ to contain more interstitial defects (Frenkel defects) than $CaF_2$ (see **Section 18.5**).  The $F^-$ ions in a perfect crystal of $PbF_2$ would occupy all of the tetrahedral interstitial sites in the close-packed $Pb^{2+}$ ion lattice, with all of the octahedral interstitial sites vacant.  Therefore, $F^-$ ions in $PbF_2$ might be expected to vacate some tetrahedral sites and to occupy some octahedral sites.  You might also be alert to the fact that lead can exhibit two stable oxidation states, $Pb^{2+}$ and $Pb^{4+}$.  However, the displacement of $F^-$ from tetrahedral to octahedral sites appears to account for most of the defects in $PbF_2$, rather than the alternative possibility of $Pb^{4+}$ occupying some $Pb^{2+}$ sites.

(c)  **$Al_2O_3$ or $Fe_2O_3$.** Both of these compounds have the corundum structure, and both $Al^{3+}$ and $Fe^{3+}$ have approximately the same ionic radius.  The important difference in this case is the possibility of variable oxidation states for iron. $Fe_2O_3$ might contain $O^{2-}$ ion vacancies that are charge compensated by the presence of some $Fe^{2+}$ ions.

**18.3 Distinguish intrinsic from extrinsic defects.** Intrinsic defects do not alter the stoichiometry of the compound. They may involve charge compensating vacancies of cations and anions or the movement of ions from their normal sites to interstitial sites. Examples are Schottky and Frenkel defects. Extrinsic defects change the elemental composition of the compound in one of two ways. Sometimes another element is added to the compound (this can be called an impurity if the new element is added accidentally, or a dopant if it is added intentionally). In other cases, extrinsic defects result from vacancies that are charge compensated by oxidation or reduction of some of the remaining ions. In these cases, no new elements are present, but the original elements are not necessarily present in the ratio of whole numbers (i.e., a nonstoichiometric compound results from this type of extrinsic defect). Examples are As-doped Si, color centers, and $VO_{0.8}$.

**18.4 Where might intercalated $Na^+$ ions reside in the $ReO_3$ structure?** The unit cell for $ReO_3$, which is shown in **Figure 18.17(a)**, is reproduced in the figure on the left below (the $O^{2-}$ ions are the open spheres). As you can see, the structure is very open, with a very large hole in the center of the cell (if the Re–O distance is $a$, the distance from an $O^{2-}$ ion to the center of the cell is $1.414a$). A $ReO_3$ unit cell containing a $Na^+$ ion (large, heavily shaded sphere) at its center is shown on the right below.

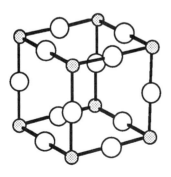

$ReO_3$

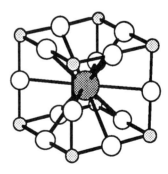

$Na_xReO_3$

**18.5    What is a crystallographic shear plane?** A crystallographic shear plane may be considered a defect or a way of describing a new structure (see the end of **Section 18.1**).   When crystallographic shear planes are distributed randomly throughout the solid, they are called Wadsley defects.    A continuous range of composition is possible, because a new, discrete phase has not been formed.   For example, tungsten oxide can have a composition ranging from $WO_3$ to $WO_{2.93}$. However, at W/O ratios higher than $1/2.93$, the shear planes are distributed in a non-random, periodic manner (i.e., a new stoichiometric phase has been formed, for example $W_{20}O_{58}$ (W/O = $1/2.90$)).   Thus, crystallographic shear planes are defects when disordered and gives a new phase with a new structure when ordered.

**18.6    Interstitial sites in the rock salt structure.** A figure showing the rock salt structure is shown on the left below.  The anions are the larger spheres, most of which are not shaded and three of which are heavily shaded.  One cation is black and the others are lightly shaded.  Since the cations in the rock salt structure occupy *all* of the octahedral interstices in the close-packed anion lattice, the only available sites for cation migration are tetrahedral interstices.   In the figure below, the black cation is shown migrating to a tetrahedral site, through "the bottleneck," a trigonal planar array of anions (the three heavily shaded spheres).   Note that there are eight equivalent trigonal planar arrays leading to eight equivalent tetrahedral sites for the migrating black cation.

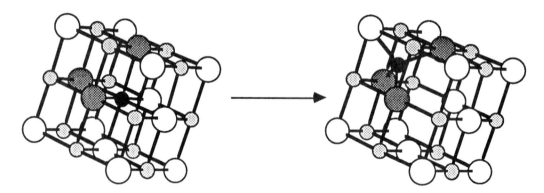

A simpler way to represent this situation is shown below.  The two close-packed layers are drawn separately and superimposed.  In the separate layers, the position of the black cation is indicated by the letter O (it is in an octahedral hole between the two layers), while the six tetrahedral holes between these two layers are indicated by the letter T. In addition to migrating to any one of the six T positions, the black cation can migrate to a tetrahedral hole above O (between the nonshaded layer of anions and a third layer) or below O (between the shaded layer of anions and a fourth layer ).

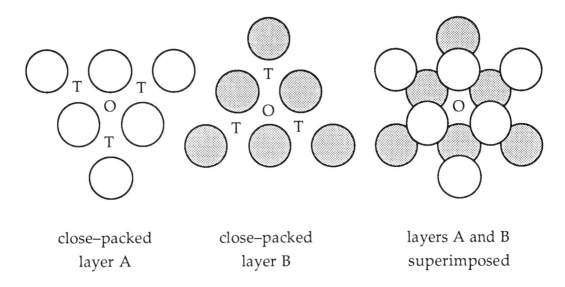

close–packed layer A      close–packed layer B      layers A and B superimposed

**18.7    Contrast the nature of the defects in TiO with those in FeO.** Titanium monoxide contains a very high concentration (~12 M) of Schottky defects (see **Section 18.1**). Remember that this type of defect results in both cation and anion vacancies. Since the stoichiometry of the compound is 1:1 in this case, there are an equal number of $Ti^{2+}$ and $O^{2-}$ vacancies. In contrast, FeO contains $Fe^{2+}$ vacancies without any $O^{2-}$ vacancies. The excess negative charge that results from the removal of some $Fe^{2+}$ ions is compensated by the formation of two $Fe^{3+}$ ions per $Fe^{2+}$ vacancy.

**18.8    How might you distinguish between a solid solution and a series of discrete crystallographic shear plane structures?** A solid solution would contain a random collection of crystallographic shear planes, whereas a series of discrete structures would contain ordered arrays of crystallographic shear planes. These two possibilities are represented in the figures below.

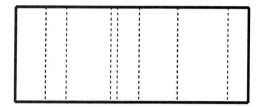

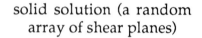

solid solution (a random
array of shear planes)

discrete phases (a series of
ordered arrays of shear planes)

Due to the lack of long range order, the solid solution would give rise to an electron micrograph showing a random distribution of shear planes. In addition, the solid solution would not give rise to new X-ray diffraction peaks. In contrast, the ordered phases of the solid on the right would be detectable by electron microscopy and by the presence of a series of new peaks in the X-ray diffraction pattern, arising from the evenly spaced shear planes.

**18.9** **Label regions in the iron-oxygen phase diagram.** An expanded portion of **Figure 18.12** is reproduced below. Only the region from 27% to 30% oxygen is shown. The vertical temperature axis runs from 500 °C to 1700 °C. There are seven regions in this portion of the iron-oxygen phase diagram. The "liquid" region is already labelled in **Figure 18.12**. At temperatures around 500 °C, the region on the left is $Fe_3O_4$ solid solution. A solid solution is a solid with a range of compositions arising from defects (see the **Further Information** section at the end of **Chapter 18**). To the right is a solid phase consisting of a physical mixture of the two crystalline phases $Fe_3O_4$ solid solution plus $Fe_2O_3$ solid solution ($Fe_3O_4$ ss/$Fe_2O_3$ ss). As discussed in the **Further Information** section, single-phase regions such as $Fe_3O_4$ ss or liquid must be separated by regions containing two phases ($Fe_3O_4$ ss plus liquid). Using this principle and **Figure 18.12**, the regions labelled "ss + l" must be a mixture of $Fe_3O_4$ ss plus liquid. Finally, since the region to the right of 30% oxygen is pure $Fe_2O_3$, the region between $Fe_3O_4$ ss/$Fe_2O_3$ ss and $Fe_2O_3$ is the pure phase $Fe_2O_3$ ss, and it is separated from the liquid phase by another two-phase region, $Fe_2O_3$ ss plus liquid.

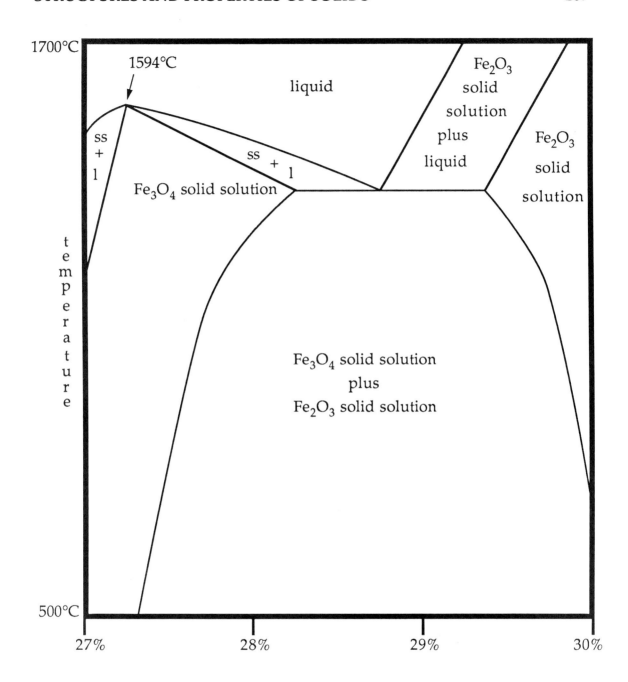

mass percentage of oxygen

**18.10 The ease of oxidation and reduction of TiO, MnO, and NiO.** The standard potentials for oxidation and reduction can be found in **Appendix 6**. Since $\Delta G° = -nFE°$, the more positive the potential (or the less negative the potential), the more favorable the process. Consider the following oxidations:

$$Ti^{2+} (aq) \rightarrow Ti^{3+} (aq) + e^- \qquad E° = 0.37 \text{ V}$$

$$Mn^{2+} (aq) \rightarrow Mn^{3+} (aq) + e^- \qquad E° = -1.5 \text{ V}$$

$$Ni^{2+} (aq) + 2H_2O \rightarrow NiO_2 (s) + 2e^- + 4H^+ (aq) \qquad E° = -1.593 \text{ V}$$

According to the criterion mentioned in the exercise, NiO will be the most difficult of these three compounds to oxidize, since $Ni^{2+}$ is the most difficult of the three aqua ions to oxidize. Now consider the following reductions:

$$Ti^{2+} (aq) + 2e^- \rightarrow Ti (s) \qquad E° = -1.63 \text{ V}$$

$$Mn^{2+} (aq) + 2e^- \rightarrow Mn (s) \qquad E° = -1.18 \text{ V}$$

$$Ni^{2+} (aq) + 2e^- \rightarrow Ni (s) \qquad E° = -0.257 \text{ V}$$

Since $Ni^{2+}$ is the easiest of the three aqua ions to reduce, NiO is likely to be the easiest of the three monoxides to reduce. In general, we expect gross similarities between redox chemistry of ions in aqueous solution and in metal oxides.

**18.11 The electrical conductivity of TiO and NiO.** TiO is a metallic conductor while NiO is a semiconductor (see **Section 18.4**). The conductivity of TiO decreases with increasing temperature, like that of a metal. The conductivity of NiO, in contrast, increases with increasing temperature. The $d$ orbitals on $Ti^{2+}$, which is early in Period 4, are relatively larger than those on $Ni^{2+}$, which is in the middle of Period 4 (see figures below). The $d$ orbitals on the $Ti^{2+}$ ions in TiO are large enough to overlap and form bands that are partly filled, a characteristic of metallic conductors. This cannot occur in NiO.

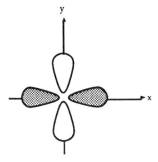

$d_{x^2-y^2}$ orbital on $Ti^{2+}$

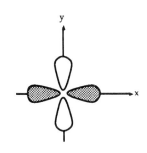

$d_{x^2-y^2}$ orbital on $Ni^{2+}$

**18.12 The Ag⁺ ion conductivity of AgI.** The diagram for this experiment is shown on the right. The shaded area between the two silver electrodes is a pellet of AgI that is heated to 165 °C. When the two electrodes are connected to a 0.1 V battery, a current passes through what has become a complete circuit. The negative pole of the battery (the left side) is a source of electrons.

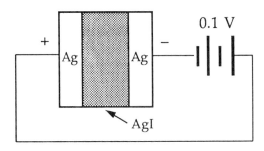

As current passes, $Ag^+$ ions in the pellet of AgI migrate to the right through the lattice and combine with electrons to form Ag atoms, which plate out on the electrode (i.e., the $Ag^+$ ions are reduced. At the electrode on the left side of the apparatus, Ag atoms are oxidized to $Ag^+$ ions and free electrons. The $Ag^+$ ions migrate through the lattice and the electrons pass through the external wire to the positive pole of the battery (the right side). The mass of the left electrode decreases while the mass of the right electrode increases.

**18.13 The changes in the silver anode and cathode in a Ag⁺ ion conductance experiment.** In the Ag/AgI/Ag cell shown above, the cathode is the silver electrode on the right side (i.e., the electrode at which reduction of $Ag^+$ ions to Ag atoms occurs). The anode is the left electrode. As current passes, Ag atoms in the left electrode are oxidized to $Ag^+$ ions, which migrate through the hot AgI to the right electrode, where they are reduced back to Ag atoms and plate out on the electrode. Therefore, over a period of time the left electrode loses mass and the right electrode gains mass. After $10^{-3}$ mole of electrons have passes through the cell, $10^{-3}$ mole of Ag atoms will have migrated from the left electrode to the right electrode. Therefore, the left electrode will have lost $(10^{-3}$ mol$)(108$ g mol$^{-1}) = 0.108$ g Ag and the right electrode will have gained 0.108 g Ag. There will not be any change in the mass of the AgI pellet.

**18.14 Contrast the mechanism of interaction between unpaired spins in a ferromagnetic material from that for an antiferromagnet.** As discussed in **Box 18.1**, in a ferromagnetic substance the spins on different metal centers (atoms or ions) are coupled into a *parallel* alignment. In the diagram on the left below, each arrow represents the spin on one metal center. Ordered regions of a sample, called domains, are shown in the bottom figure. In an antiferromagnetic substance, the spins on different metal centers are coupled into an *antiparallel* alignment. As the temperature approaches absolute zero, the net magnetic moment of a ferromagnetic domain becomes very large, while that of an antiferromagnetic domain goes to zero.

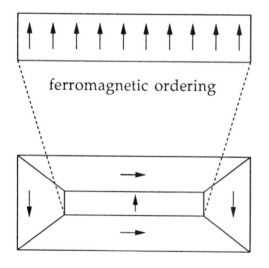

ferromagnetic ordering

antiferromagnetic ordering

One possible domain structure for a ferromagnetic crystal. The arrows represent the net magnetic moment for each domain. The crystal as a whole has a net magnetic moment.

The direction of the net moment =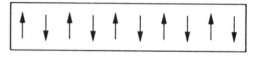

**18.15 Suggest a distribution of cations between octahedral and tetrahedral sites for $CoFe_2O_4$.** This compound is an example of an $AB_2O_4$ oxide. These compounds either exhibit the spinel ($MgAl_2O_4$) or inverse spinel (e.g., $Fe_3O_4$) structure. When paramagnetic ions are present, both types of structures exhibit antiferromagnetism — the tetrahedral ions have their spins aligned antiparallel to the spins of the octahedral ions. The cations in $CoFe_2O_4$ are $Co^{2+}$ ($d^7$) and $Fe^{3+}$ ($d^5$). In the spinel structure, the +2 cations occupy tetrahedral sites and the +3 cations occupy octahedral sites. If $CoFe_2O_4$ exhibited the spinel structure, there would be 7 spins per formula unit (3 spins "up" from one $Co^{2+}$ ion and 10 spins down" from two $Fe^{3+}$ ions). In the inverse spinel structure, half of the +3 ions occupy tetrahedral sites and the other ions occupy octahedral sites. If $CoFe_2O_4$ exhibited the inverse spinel structure, there would be 3 spins per formula unit (5 spins "up" from one $Fe^{3+}$ ion and 8 spins "down" from one $Co^{2+}$ ion and one $Fe^{3+}$ ion. Since the observed number of spins is 3.4, it appears as though $CoFe_2O_4$ has the inverse spinel structure. Using the notation introduced in **Section 18.5**, we could write the formula of $CoFe_2O_4$ as $Fe[CoFe]O_4$ (recall that the square brackets are used to denote the ions occupying octahedral sites).

**18.16 Describe the difference between the structure of $YBa_2Cu_3O_7$ and perovskite.** The perovskite structure is exhibited by many $ABO_3$ oxides (the prototype is $CaTiO_3$). It is shown in **Figure 18.18** and is reproduced below.

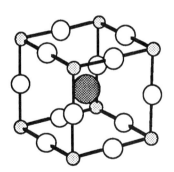

The structure of perovskite, $CaTiO_3$. The open spheres are $O^{2-}$ ions, the small shaded spheres are $Ti^{4+}$ ions, and the large shaded sphere in the center of the cell is the $Ca^{2+}$ ion.

The structure of the high-$T_c$ superconductor $YBa_2Cu_3O_7$ is related to the peroxskite structure in the following way. The A ions are $Y^{3+}$ *and* $Ba^{2+}$ ions. The B ions are copper ions, which have an average oxidation state of +2.33. Thus, the 1:1 cation stoichiometry of perovskite is retained in $YBa_2Cu_3O_7$. However, if $YBa_2Cu_3O_7$ had precisely the same structure as perovskite, there would have to be nine $O^{2-}$ ions per formula unit, not seven (the formula unit would be $(Y_{1/3}Ba_{2/3}CuO_3)_3 = YBa_2Cu_3O_9 \neq YBa_2Cu_3O_7$). Therefore, this superconductor is missing some $O^{2-}$ ions per formula unit relative to perovskite. For this reason, none of the copper ions are six-coordinate. Instead, some are five-coordinate and some are four-coordinate. Furthermore, the $Y^{3+}$ and $Ba^{2+}$ ions do not have the twelve-coordinate dodecahedral coordination of the $Ca^{2+}$ ions in perovskite.

# CHAPTER 19

# BIOINORGANIC CHEMISTRY

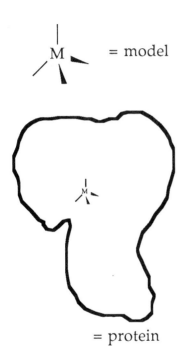

= model

= protein

Bioinorganic chemistry is concerned with the roles of metal ions in biological systems. Many of the spectroscopic tools that inorganic chemists use to study the kinds of complexes discussed in earlier chapters can be used to study metalloproteins and other metal-containing biomolecules. However, the structure of the active sites in proteins cannot be determined very accurately using X-ray diffraction. Therefore, bioinorganic chemists often study the structures of low molecular weight "model" complexes and make the following inference: if there is a spectral congruence between the model and the protein, it is assumed that there is a structural congruence as well.

## SOLUTIONS TO IN-CHAPTER EXERCISES

**19.1   Comment on the extent of Co to $O_2$ electron transfer.** The oxygen species present in $KO_2$ and $BaO_2$ are superoxide ion ($O_2^-$) and peroxide ion

$(O_2^{2-})$. The O–O bond distance in $[Co(CN)_5O_2]^{3-}$ is just a little longer than the distance in $O_2$, so very little electron transfer is evident. The O–O distances in $[Co(bzacen)(py)O_2]$ and $[(NH_3)_5CoO_2Co(NH_3)_5]^{5+}$ are similar to the distance in $KO_2$, so they are thought to contain an oxygen species similar in electronic structure to superoxide ion. The O–O distance in $[(NH_3)_5CoO_2Co(NH_3)_5]^{4+}$ is nearly the same as in $BaO_2$, so the oxygen species is inferred to be similar in electronic structure to peroxide ion. If you calculate the apparent oxidation state of cobalt in each of these four complexes assuming that the above charges on the oxygen species are correct, you will find that it is Co(II) in $[Co(CN)_5O_2]^{3-}$ and Co(III) for the other three.

**19.2   Propose a hydroxo mechanism for carbonic anhydrase.** If a coordinated water molecule is activated for nucleophilic attack, there should be a basic amino acid sidechain nearby. This will help stabilize the metal bound –OH group and enhance its nucleophilicity. Carbon dioxide is probably hydrogen bonded at one or both ends in order to orient it for nucleophilic attack and to stabilize the negative charge that will build up on the $CO_2$ oxygen atoms. A possible mechanism is shown on the right (–B: = a basic amino acid sidechain). The final steps (not shown) would be (i) dissociation of $HCO_3^-$ away from the active site, (ii) removal of $H^+$ from the –BH group by solvent, and (iii), coordination of a solvent water molecule to the vacant site on the $Zn^{2+}$ ion.

**19.3 Testing the Marcus theory.** Let us construct an abbreviated table with the data that we need:

| entry | $E°$ (V) | distance (Å) | ratio of $k_r$'s |
|---|---|---|---|
| first | 0.2 | 8 | |
| sixth | 0.2 | 10 | 600 |
| third | 0.4 | 8 | |
| eighth | 0.4 | 16 | 40,000 |
| fourth | 1.1 | 8 | |
| ninth | 1.0 | 16 | 22 |

The two members of the three pairs of entries have the same net potential. In all three cases, the redox couple with the shorter distance has the faster rate, consistent with the argument that the probability of finding the electron on the other side of a high barrier decreases with increasing width of the barrier. If you compare the first pair of entries with the second pair, you can see that a doubling of the distance produces a much larger decrease in the rate constant than simply changing the distance from 8 Å to 10 Å. However, the third pair of entries shows that the distance change is not the only parameter that is important. A doubling of the distance here produces a relatively small change in $k_r$, so other factors must come into play.

**19.4 The electron configuration of the resting state of P-450.** The Fe(III) oxidation state corresponds to a $d^5$ configuration. Most five-coordinate Fe(III) porphyrin complexes are high-spin, not low spin. Therefore, each $d$ orbital has one electron. The $d$-orbital splitting pattern for a $C_{4v}$ complex is shown in **Figure 14.5**. For the resting state of P-450, it would look like the figure on the right.

$$\text{—↑— } b_1 \ (x^2\text{–}y^2)$$

$$\text{—↑— } a_1 \ (z^2)$$

$$\text{↑↓↑— } b_2 \ (xy)$$
$$\text{↑↓ ↑↓ } e \ (xz, yz)$$

$$C_{4v} \ Fe^{3+}$$

**19.5 Difficulties in the preparation of $[Fe_2S_2(SR)_4]^{2-}$.** The oxidation state of the two Fe atoms in this dinuclear cluster is $Fe^{3+}$, whereas the average oxidation state of the four Fe atoms in $[Fe_4S_4(SR)_4]^{2-}$ is $Fe^{2.5+}$. Therefore, the difficulty is that the $Fe^{3+}$ is too oxidizing for a $SR^-$ ligand (see the answer to the **Example**). The conversion of $[Fe_2S_2(SR)_4]^{2-}$ to $[Fe_4S_4(SR)_4]^{2-}$ occurs by oxidation of $SR^-$ to RSSR and reduction of $Fe^{3+}$ to "$Fe^{2.5+}$," as follows:

$$2[Fe_2S_2(SR)_4]^{2-} \rightarrow [Fe_4S_4(SR)_4]^{2-} + 2RSSR$$

Notice how each of the six faces of the $Fe_4S_4$ core of $[Fe_4S_4(SR)_4]^{2-}$ (see **Structural Drawing 26**) is similar to the $Fe_2S_2$ core of $[Fe_2S_2(SR)_4]^{2-}$ (see **Structural Drawing 25**).

| SOLUTIONS TO END-OF-CHAPTER EXERCISES |
|---|

**19.1 Bond length and net spin for $O_2$, $O_2^-$, and $O_2^{2-}$ as ligands.** According to **Figure 2.22**, $O_2$, $O_2^-$, and $O_2^{2-}$ have 2, 3, and 4 electrons, respectively, in the $2\pi_g$ set of antibonding orbitals. $O_2$, with the fewest antibonding electrons, will have the strongest and shortest bond, while $O_2^{2-}$, with the greatest number of antibonding electrons, will have the weakest and longest bond. With two unpaired electrons, $O_2$ will have a net spin of 1. With one unpaired electron and no unpaired electrons, respectively, $O_2^-$ and $O_2^{2-}$ will have net spins of 1/2 and 0, respectively.

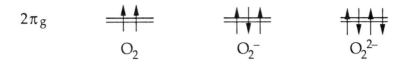

$2\pi_g$      $O_2$      $O_2^-$      $O_2^{2-}$

**19.2  Oxygen affinity of hemoglobin and myoglobin.** The net reaction Hb + Hb(O$_2$)$_4$ $\rightleftharpoons$ 2Hb(O$_2$)$_2$ is the sum of the two equilibria:

$$Hb + 2O_2 \rightleftharpoons Hb(O_2)_2$$

$$Hb(O_2)_4 \rightleftharpoons Hb(O_2)_2 + 2O_2$$

The first equilibrium lies to the left (i.e., hemoglobin that has no O$_2$ bound to it has a relatively low affinity for O$_2$). The second equilibrium also lies to the left (in this case, consider the reverse reaction — hemoglobin that has some O$_2$ bound to it has a high affinity for O$_2$). Therefore, the net reaction lies to the left. The position of the equilibrium Hb(O$_2$)$_4$ + 4Mb $\rightleftharpoons$ Hb + 4Mb(O$_2$) depends on the partial pressure of O$_2$. At low partial pressures, where the Hb curve is well below the Mb curve (see **Figure 19.7**), this reaction lies to the right (myoglobin has a much higher affinity for O$_2$ than hemoglobin at low partial pressures). At high partial pressures of O$_2$, the affinity of hemoglobin for O$_2$ approaches that of myoglobin, and the reaction lies farther to the left.

**19.3  A mechanism for CO poisoning.** Since O$_2$ and CO are similar electroni-cally as well as sterically, they can both bind to the same sites in metallo-proteins such as hemoglobin. In fact, hemoglobin and myoglobin (and model iron porphyrin complexes) have a higher affinity for CO than for O$_2$. Therefore, if CO is present, it can bind to, and block, the oxygen transport sites of hemoglobin, thus preventing oxygen from being distributed to various tissues.

**19.4  Peptide synthesis at Co(III) centers.** The ester carbonyl C atom is the electrophilic center that is subject to nucleophilic attack. Coordination of the carbonyl O atom to Co(III) increases the partial positive charge on this C atom, enhancing its electrophilicity. The nucleophile is the amino N atom of the peptide being added. The leaving group is the ester $OCH_3^-$ group, which is probably protonated before the C–O bond cleavage occurs.

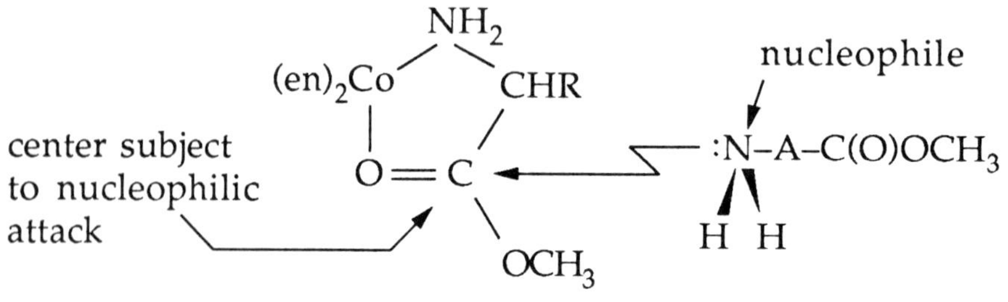

center subject to nucleophilic attack

$$(en)_2Co \overset{NH_2}{\diagup} \overset{\diagdown}{\underset{O=C}{\diagup}} CHR$$

nucleophile

:N–A–C(O)OCH$_3$

OCH$_3$

H  H

**19.5   Co(II) substitution for Zn(II).**  The spectral differences between Co(II) and Zn(II) are a consequence of their $d$-electron configurations — $d^7$ for Co(II) and $d^{10}$ for Zn(II).  Regardless of geometry, Co(II) complexes always have one or three unpaired electrons (usually three) and can be studied by electron paramagnetic resonance spectroscopy (EPR, see **Section 14.9**).  A $d^{10}$ metal ion such as Zn(II) cannot have unpaired electrons in any geometry, and never exhibits an EPR spectrum (Zn(II) complexes are said to be "EPR silent").  You should recall that EPR can be used to map the extent to which the unpaired electrons are delocalized onto the ligands.  With an incomplete $d$ subshell, Co(II) complexes can undergo ligand field ($d$-$d$) transitions that result in visible absorption spectra (see **Section 14.3**).  In contrast, Zn(II) complexes are generally colorless (charge transfer transitions can render some Zn(II) complexes colored, but this is rare).  Finally, since Co(II) complexes are colored, they can also be studied by circular dichroism spectroscopy (see **Section 14.8**).  The assumption that must be made when studying Co(II) substituted zinc proteins is that isomorphous substitution has taken place.  This means that the coordination number and geometry of the Zn(II) ion in the native protein has not changed upon Co(II) substitution.

**19.6   High-spin and low-spin Fe(II) porphyrin complexes.**  The electron configurations for the two spin states in question are $t_{2g}^4 e_g^2$ (high-spin) and $t_{2g}^6$ (low-spin), as shown below:

high–spin
Fe$^{2+}$

$e_g$

$t_{2g}$

low–spin
Fe$^{2+}$

High-spin Fe(II) is the larger of the two because of the two $e_g$ electrons that it possesses. **Section 7.4** described the metal-ligand bonding in octahedral complexes. The metal $t_{2g}$ orbitals are either nonbonding or weakly $\pi$-bonding MOs, but the metal $e_g$ orbitals are always metal-ligand $\sigma$-antibonding MOs (see **Figures 7.7** and **7.10**). Population of $\sigma$-antibonding MOs leads to longer metal-ligand bond distances and an apparently larger metal ion radius. Only weak ligands such as $H_2O$ and ethers induce a small enough $\Delta_o$ to yield six-coordinate high-spin Fe(II) porphyrin complexes (one that has been studied by X-ray crystallography is [Fe(porphyrin)(thf)$_2$]). More strongly basic ligands such as pyridine (py) and the imidazole group of histidine, or $\pi$-acid ligands such as CO, yield low-spin complexes (a structurally characterized one is [Fe(porphyrin)(py)(CO)]).

**19.7   Why are $d$ metals such as Mn, Fe, Co, and Cu used in redox enzymes in preference to Zn, Ga, and Ca?** The first group of metals occur naturally in redox enzymes because they each have at least two stable oxidation states. Redox catalysis involves the cyclic oxidation and reduction of the metal ion (for example, from Fe(II) to Fe(III) and back again for the cytochromes). The metals in the second group have only one stable oxidation state (Zn(II), Ga(III), and Ca(II)), so they cannot be oxidized or reduced at physiological potentials (i.e., the potentials of the most oxidizing and most reducing proteins in a cell).

**19.8   Why is it difficult to measure the self-exchange rate constants of redox enzymes?** A redox enzyme has been "designed" to bind a substrate molecule or ion and to transfer electrons to it. The binding site is usually highly specific for the substrate. Furthermore, once bound, substrate is held in close proximity to the metal center. The self-exchange rate constant is defined as the rate constant for electron transfer between the oxidized form of the enzyme and the reduced form. In the physiological redox reaction, the two forms of the redox enzyme do not transfer electrons with one another. In fact, the self-exchange electron transfer may be very slow, even if the enzyme-substrate electron transfer is fast. Therefore, since the site(s) at which the two forms of the enzyme bind to each other may not be the same as (or even close to) the substrate binding site, the self-exchange rate constant for a redox enzyme is almost meaningless.

**19.9 The specificity of protein-protein outer-sphere electron transfer.** One way that the protein parts of an enzyme contribute to making outer-sphere electron transfer highly specific is by controlling the "docking" of the enzyme with its redox partner. This controls the distance between the metal centers as well as their relative orientations, as illustrated in the figure below. Another way is by holding the ligands for each metal center in a particular geometry, thereby controlling inner-sphere reorganization energies.

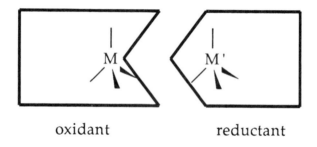

oxidant                    reductant

**19.10 Identify one significant role in biological processes for the elements Fe, Mn, Mo, and Zn.** See Table 19.4:

Fe:   oxygen transport (hemoglobin, hemerythrin)
       oxygen storage in tissue (myoglobin)
       electron transfer (cytochromes, rubredoxin, ferredoxins)
       oxygenation of hydrocarbons (cytochrome P-450)

Mn:  oxidation of water to $O_2$ (photosystem II)

Mo:  reduction of $N_2$ to $NH_3$, or nitrogen fixation (nitrogenase)

Cu:  oxygen transport (hemocyanin)
     electron transfer (blue copper proteins)

Zn:  conversion of $CO_2$ to $HCO_3^-$ (carbonic anhydrase)
     hydrolysis of peptide linkages (carboxypeptidase)
     decarboxylation of oxaloacetic acid (oxaloacetate decarboxylase)

**19.11 What prevents simple iron porphyrins from functioning as $O_2$ carriers?**  Simple iron porphyrins are not protected from aggregation as are the iron porphyrin prosthetic groups in hemoglobin and myoglobin. Aggregation of the two complexes [Fe(porphyrin)$O_2$] and [Fe(porphyrin)] leads to a μ-peroxo dimeric complex, [[Fe(porphyrin)]$_2O_2$], and ultimately to a μ-oxo dimeric complex, [[Fe(porphyrin)]$_2$O]. Note that the oxidation state of iron is Fe(II) in [Fe(porphyrin)] and Fe(III) in [[Fe(porphyrin)]$_2$O]. Thus, treatment of simple [Fe(porphyrin)] complexes with $O_2$ leads to irreversible *oxidation* instead of reversible *oxygenation*.

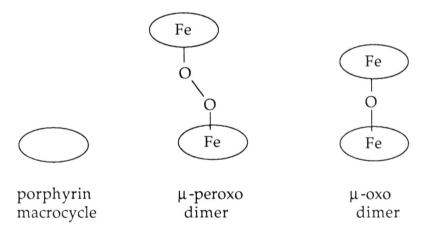

porphyrin        μ-peroxo        μ-oxo
macrocycle        dimer          dimer

**19.12 Sketch the steps illustrating a metal complex functioning as each of the following: (a) a Brønsted acid in an enzyme-catalyzed reaction.**  A water molecule coordinated to a metal ion is a stronger Brønsted acid than pure water (see **Section 5.5**). The functioning of carbonic anhydrase is believed to involve a Zn–$OH_2$ species (see the solution to **In-Chapter Exercise 19.2**).

**(b) A Lewis acid in an enzyme-catalyzed reaction.** The coordination of a basic ligand to an electrophilic metal ion induces a positive charge at the atom adjacent to the point of coordination. For example, coordination of a carbonyl oxygen atom to Zn(II) enhances the partial positive charge at the carbonyl carbon atom. The functioning of carboxypeptidase is believed to involve a Zn–O=C(N–)(R) species (see **Figure 19.11**).

**19.13  Determine the average oxidation state of iron in $[Fe_4S_4(SR)_4]^{2-}$.** Each sulfide ion is $S^{2-}$ and each thiolate ligand is $SR^-$. Four of each of these ligands results in a total ligand charge of 12–. Since the tetranuclear complex has a 2– charge, the four iron atoms must share a 10+ charge. Therefore, the average oxidation state of iron is $Fe^{2.5+}$.

**19.14  What is the average change in oxidation number of the Mn atoms in the $S_4$ to $S_0$ step in the PS II cycle?** The oxygen atoms in the $S_4$ (adamantane) $Mn_4O_6$ complex (see **Figure 19.20**) and the $S_0$ (cubane) $Mn_4O_4$ complex are formally $O^{2-}$ ions. The half-reaction involving oxygen evolution must be $2O^{2-} \rightarrow O_2 + 4e^-$. Therefore, since the four Mn atoms are reduced by a total of four electrons, each Mn atom experiences a change of –1 in oxidation number.

**19.15  Could magnesium be used as the metal ion in a cytochrome?** In the case of chlorophyll, the source of the electron is a macrocycle $\pi^*$ orbital (it is promoted to this orbital from the macrocycle HOMO, a $\pi$ orbital, by absorption of a photon). The Mg atom is involved in the spatial orientation of the chlorophyll molecule relative to other parts of the reaction center, but is not itself involved in electron transfer. In contrast, the source of the electron in a cytochrome is the iron atom itself, as it changes oxidation state from Fe(II) to Fe(III). Magnesium could not be effective in a cytochrome, since Mg(II) cannot be oxidized to a higher oxidation state by any chemical oxidant.